Islands of New Brunswick

Living Between the Tides

Allison Mitcham
Illustrations by Peter Mitcham

Nimbus Publishing Limited
3731 Mackintosh St, Halifax, NS B3K 5A5
(902) 455-4286 nimbus.ca

Printed and bound in Canada

Design: Jenn Embree
Cover Illustration: Jenn Embree
Author Photo: Peter Mitcham

Library and Archives Canada Cataloguing in Publication

Mitcham, Allison, 1932-
Islands of New Brunswick : lives at low tide / Allison Mitcham.
Includes bibliographical references and index.
Issued also in electronic formats.
ISBN 978-1-77108-022-4

1. Islands--New Brunswick—History. 2. New Brunswick—History. I. Title.

FC2461.M58 2013 971.5'1 C2012-907383-0

Canada
NOVA SCOTIA Communities, Culture and Heritage
Canada Council for the Arts Conseil des arts du Canada

Nimbus Publishing acknowledges the financial support for its publishing activities from the Government of Canada through the Canada Book Fund (CBF) and the Canada Council for the Arts, and from the Province of Nova Scotia through the Department of Communities, Culture and Heritage.

FSC www.fsc.org MIX Paper from responsible sources FSC® C107923

Contents

Acknowledgements

I wish to thank the research council of the Université de Moncton for assistance given in research for this book.

In addition, I wish to thank the following people for their information and moral support: Chief and Mrs. Peter Barlow, Mr. and Mrs. Wainwright Weston, Father Charles Mercereau, Marvin Snowdon, Noel and Jean Justason, Bernard Holt, the Saint Andrews Historical Society; my editor at Nimbus Publishing, Whitney Moran; my daughter, Stephanie, who accompanied me to many of these offshore islands and who has long been my most helpful support and critic for my books; and my husband, Peter, for his lovely illustrations.

Introduction

Like ships, small islands are fascinating—and for many of the same reasons. Surrounded by water and cut off from conventional society on the mainland, life is strictly governed by storms and tides as well as by the unavoidable interactions of human beings confined to a limited space. On a ship as on an island, it is virtually impossible to avoid close contact with the others who live there too; and so, as countless writers of sea stories have shown, this makes for different sorts of relationships between individuals. Depending on circumstances, both tolerance and hostility have a deeper hold on ships and islands than on the mainland, where individuals are freer to come and go independently of their neighbours. After all, one can just as easily leave some of our offshore islands in a storm or fog or shifting ice as one can put out from an ocean liner in the mid-Atlantic. Because of this, dramas at sea and on islands develop, whereas on land the tensions causing them are more apt to be diffused.

A good example of explosive tension resulting in a dramatic series of events occurred early in the nineteenth century on Grand

Manan, when Mr. Wilfred Fisher, nicknamed "The Emperor of Grand Manan," took the law into his own hands because he felt that his power was threatened. Like the captain of a ship, he apparently demanded a sort of absolute allegiance from the islanders—an allegiance unthinkable on the mainland. Like a power-mad captain in a Melville novel, Mr. Fisher set out to demonstrate his authority, convinced that he was the law. (The events leading up to Mr. Fisher's actions and the outcome are discussed in the chapter on Grand Manan.)

Drama is also triggered on both ships and islands by accidents and emergencies that can be more easily dealt with on the mainland because of access to special equipment and personnel. Fires, storms, sudden and severe illnesses, accidents—these all seem more traumatic occurrences at sea than on the mainland. At the same time, such events also call for the sort of hardiness and individual initiative that is less often needed on the mainland.

For these reasons, islands and ships are particularly rich sources of folklore. There is a mystique surrounding both—an aura of romance, mystery, and adventure. Tales of ghosts, pirates, treasure, strange and inexplicable happenings, heroism, and disaster hover with the persistence of seagulls. There is a uniqueness to each boat and island. Aside from obviously having certain basics in common, any long-time sailor or islander would be quick to dispel a mainlander's contention that any one ship or island is much like another; they are individual, manned by crews of very different sorts.

The islands off the coasts of and New Brunswick that are represented in this book have been selected because they are so different from one another, and because each is so special in its own way. The islands vary in size, geographical settings, topography, the ethnic and religious backgrounds of their inhabitants, and in historical, ecological, aesthetic, and commercial significance. Although all have at one time been connected with fishing, each has other associations. Some are important as bird nesting grounds; others have so strongly grasped the imagination of visitors that they have been the settings of novels and poems; and most are associated with exceptional individuals—some internationally famous, others well-known in Atlantic Canada, and still others who remained unknown before the appearance of this book. Finally, the connection of each of these islands to the nearby mainland is, to some extent, important here. There are also links among the islands—the Fundy and Northumberland Strait islands for instance. Yet, while recognizing the regional similarities that certain islands share, it is nevertheless the uniqueness of each island that remains the dominant and lasting impression.

Grand Manan Island in the Bay of Fundy

Grand Manan

The name MANAN, of Grand and Petit Manan, is certainly of Maliseet-Passamaquoddy-Penobscot Indian origin, and means simply THE ISLAND, the prefixes having been added by the French to distinguish the two.

–W. F. Ganong, *The Origin and History of the Name Grand Manan*

For the Loyalist settlers and their descendants, as for the Aboriginal people before them, Grand Manan has always been the island dominating its own archipelago as well as much of the rest of the Bay of Fundy. Whatever the difficulties of living here—and there are many (the icy water, the dangerous shoals and currents surrounding the island, the fierce gales, the dense fogs, and the almost two-hour ferry crossing from the New Brunswick mainland)—Mananers seem to almost unanimously agree that their island is the best place in the world to live.

Part of the reason for this attitude is the success of the island's fisheries, which are among the world's most flourishing. The extreme coldness of the water, the high tides and dangerous eddies, the shoals and ledges—all seem to provide just the kind of environment that the fish crave. Catches off the island have been legendary, so it is almost impossible to impress a Mananer with a fish story—he is sure to be able to top it. If you do not believe him, he may refer you to a statistical report—although not necessarily a contemporary one—to back him up. Here is one such account taken from "The Perley Report on the Fisheries of Grand Manan" (1850):

> The upper part of Grand Harbour...abounds with lobster, which during the season may be taken with a gaff, in almost any quantity.... As many as were required for the hooker were quickly taken in this way, in two to four feet of water; the places resorted to by lobsters were easily known by the holes made by them in the flats.... [1]

Word of such plenty has always spread quickly, and in the mid-nineteenth century, a state of almost open war existed between the islanders and those they considered intruders, many of whom were the Mananers' American cousins who were keen to share the bounty. Since Grand Manan is only about ten kilometres off the American coast and almost four times that distance from the Canadian mainland, the American intrusion was easy and Canadian protection was hard to come by. Still, as Perley shows in the following excerpt from his report, the Mananers were prepared to defend their interests—and often used rather unconventional methods:

> In September (1850) the number of fishing vessels at the Southern Head amounted to one hundred or more.... The presence of the Revenue Cutter alone prevented a scene of disorder and confusion, as well as great destruction of nets and other valuable property. After the nets were set for the night, all the fishing boats were ordered to return to the vessels to which they belonged; while the boats of the Cutter rowed guard during the night to prevent persons from injuring or stealing the nets. Yet, notwithstanding these precautions, and the exercise of great vigilance, nets were continually destroyed or stolen, especially during dark and windy nights, when the depredators could not be seen or

> heard. It was said, that boats with old scythes attached to their bottoms had been rowed swiftly among the nets, by which great damage had been done. [2]

Today the Mananers are better protected from poachers, and although they have to go farther afield for lobster and other fish, the abundance of their catch seems not to have diminished greatly in more than a century. "Everyone's working on the island," proud Mananers boast. This fact is, of course, no small reason for the islanders' satisfaction. Nevertheless, the men who fish such waters need to possess much skill and endurance to simply survive, let alone bring in the enormous catches they have obtained over the years. Accidents have happened, but less often than one would suppose; they have usually involved strangers to the Grand Manan archipelago. In fact, when one middle-aged woman was asked if she worried about her husband when he was fishing, she replied at once in gently reproving tones: "Oh, no, not at all! He's been fishing these waters all his life. He knows what he's doing."

It has been said over the years that the island fishermen know the waters surrounding their island so well that even if a fog descends on the fishing grounds, they can usually make home without much difficulty, as they are able to recognize each rock and ledge by the kind of rush or gurgle that the water around it emits. Their senses seem honed as precisely as an acute and sophisticated radar system. Consequently, it is hardly surprising that over the years, and particularly in the days of sailing ships, Grand Manan seamen were much in demand for work on vessels plying both American and Canadian waters and above all, on ships entering the Bay of Fundy. Sailors reckoned that no one had a better understanding of the sea at its most upset than Grand Mananers did; they were raised from infancy to interpret its most discordant music.

Neither does anyone have a stronger sense of the past and its influence on the shaping of the present than a Mananer. And nowhere else does one meet a whole population so up-to-date on its own history and so concerned with preserving it. The thriving museum and historical society attest to this as well as the fact that almost any Mananer, even if not at all bookish, seems to be imbued with an encyclopedic knowledge of his island, past and present. One of those responsible for the preservation of Grand Manan's heritage is L. W. Ingersoll, long-time editor of the excellent and unusual journal, the *Grand Manan Historian*. The following excerpt is from one of his many articles and is indicative of his vast knowledge of his subject, his flair for narrative, and his recognition of the islanders' spirit of independence—ultimately showcasing his realization that Grand Mananers have always marched to a different drummer:

> Sheep were kept the year round on Kent, Outer Wood and some of the smaller islands of the archipelago, requiring two roundups, one in the Spring for shearing, and one later in the year for culling market lambs and mutton stock. Cattle, mostly young steers to fatten for winter beef, were pastured in the same places for about six months, requiring a roundup in the Fall.
>
> Young boys were hired for a few pennies a day to help with the roundups, usually just a case of driving docile animals into a corral. Often, however, as happened when the writer was involved on two or three occasions as a young lad, a mavarick would break away and elude its pursuers. Someone had liberated domestic rabbits on Outer Wood Island and these had multiplied ten thousand fold.... Chasing a half wild steer over the crags and through the raspberry bushes with herring gulls screaming overhead in the fog, rabbits flushed and running for cover, with the risk of tripping in a burrow of a rabbit warren, or that of a Mother Carey's Chicken (Leach's petrel)—hooting and yelling to scare the steer in the right direction, was all heady stuff hardly matched by today's western classics on television.
>
> More than that, it must not be forgotten that this whole process of natural husbandry of sheep and cattle added an additional

> dimension to the security and benefits of life in this part of the world, too often derided by the well-meaning but grossly uninformed 'experts' on urban advantages. [3]

Although scores of knowledgeable and enthusiastic writers for the *Grand Manan Historian* have documented island life over the years, a flair for oral narrative is very evident among Grand Mananers in general. For instance, while crossing over to the mainland—particularly out of season—it would be very likely to overhear an islander instructing his visiting relatives on the history and geography of his island as the boat passes various landmarks: Whale Cove, Ashburton Head, and others. On nearing Ashburton Head, he would describe, as vividly as if he had been there, the wrecking of the *Lord Ashburton* on the night of January 18, 1857, at the rocky promontory on the north end of the island that now bears the name of that ill-fated barque. The islander is also bound to emphasize the fate of one of the ship's eight survivors, a Danish seaman called James Lawson who scaled the cliffs and made his way to a barn where he was found the following morning still alive, though suffering severe frostbite. After being nursed for several days by the island people, he was transferred to the marine hospital in Saint John where he remained for more than five years. During most of his long recovery, Lawson's thoughts must have continued to return to the people of Grand Manan who had saved his life, because as soon as he could move about he decided to return to the island and spend the rest of his life among its inhabitants. He married an island girl, and despite the fact that he was permanently crippled after the shipwreck, with both feet partially amputated, he learned the trades of harness maker and shoemaker. His descendants still live on the island.

Because such stories have been treasured and passed down from one generation to another, nearly every Mananer can trace his roots and those of his neighbours, knowing that they are all firmly anchored in the soil and rock of the island. Each individual who has been blown by chance or destiny into this archipelago, who took root and thrived here, seems to have found his own particular and accepted niche over the years. These islanders have looked out for each other. This is, of course, a basic survival rule of any remote society. Still, it is interesting to note that despite all the intrusions of the twenty-first century, Mananers stick together. This is apparent everywhere; but perhaps nowhere more amusingly so than on the bulletin board on the ferry, where islanders were once advised that the Island View Motel in Saint John offered Mananers two dollars off the regular price of accommodations. Such extended loyalties exist only in a very stable population, and the population of Grand Manan (about 2,400) has altered little in numbers, in background, or in occupation over the last hundred years. In each generation, some of the young people go away to make their lives elsewhere, but enough remain to carry on.

Despite the Grand Mananers' strong ties to the past, they have rarely been reluctant to accept contemporary innovations that they hoped would benefit their community. They have always been surprisingly up-to-date in the services and devices that they have imported. At no time was this more evident than in the 1880s. For instance, it was during this time that so many communities had no newspaper at all; Grand Manan had two. Both editors were exceptional and controversial figures—as well as declared rivals. Dr. G. N. Noyes ran the *Island Press* and J. B. Lorimer the *Island News*. Lorimer had already made a name for himself as a mainland journalist and the author of the still fascinating and informative book, *History of the Islands and Islets in the Bay of Fundy* (1876).

In 1881 Grand Manan established its own telegraph office. The islanders installed this service only seven years after the first successful cable across the Atlantic. Amusingly, as old records testify, Mananers most often used the telegraph system for getting in touch with neighbours and family on the island rather than as a method of making contact with the outside world. With this new equipment, family members who had gone visiting in the neighbouring village could be checked on and called home if needed, and the doctor could at last be reached quickly and easily. Dr. J. F. MacCaulay's card, issued early in the last century, announced: "Calls answered any time of Day or Night to any part of Grand Manan. Can be called by telegraph up to 9:00 P.M."

Another step forward for Mananers in the 1880s was a regular ferry service. Subsidized by both the federal and provincial governments, the first ferry went into service on July 1, 1884. Since, just prior to this, lighthouses and fog alarms had been established at such treacherous places as Gannet Rock, Machias Seal Island, Long's Eddy, and the Swallow Tail, the time seemed right for a ferry service.

All of these developments paved the way for Grand Manan's acquisition of a telephone system in the following decade—only sixteen years after Alexander Graham Bell had patented the device. By this time, Mananers felt justified in claiming that they were just as up-to-date as people anywhere, and yet not plagued by the usual unpleasant consequences of progress. This is essentially true even today, for despite the islanders' ready acceptance of modern innovations, Grand Manan has not yet been spoiled by pollution—a price that so many other once-beautiful places have had to pay for modernization and affluence. Although now threatened by potential radiation leaks from the Point Lepreau Nuclear Generating Station and oil spills from passing tankers, Grand Manan has been spared such horrors thus far.

Grand Manan, and indeed the whole surrounding archipelago, is still recognized as the special habitat and sanctuary of many species of marine and bird life, as well as being one of the world's most beautiful spots. Whales and dolphins cavort offshore—often accompanying the ferry on parts of its run—and birds nest by the hundreds of thousands on the outer, and now uninhabited islands.[4] Just as Audubon came to Whitehead Island in 1833 to observe and draw the herring gull, so do contemporary scientists and wildlife enthusiasts flock annually to Machias Seal Island and other rewarding parts of this fertile domain to make their observations.

Squalid, makeshift dwellings rarely mar the Grand Manan landscape. On the eastern and inhabited side of the island, sturdy well-kept houses stand in green fields overlooking the sea. These fields, which once provided pasture for the islanders' livestock and food for their tables, are mostly untended now. Mananers are too busy fishing to bother with farming. But the abandoned pastures and fields are not without their charm. Here, within sight of the sea, one may ramble through June meadows where, on low ground, irises flame blue amid marsh grasses and, on the uplands, strawberry blossoms gleam white against the dark earth, bright and numberless as stars in the Milky Way. Overhead, gulls shriek their delight in the clear salt air and, on the bay below, sea ducks bob and murmur complacently among mats of seaweed and moored dories.

Behind the meadows, the woods—now mostly evergreen—begin. These woods stretch almost uninterrupted for eight or nine kilometres—depending on the island's varying width—to the formidable and uninhabited western cliffs. It was in these forests, and often several kilometres from the sea, that Mananers-of-old built hundreds of boats of all sizes. Oaks and pines did not, understandably, survive this industry; nor did boat building, like farming, survive the successful expansion of the island's fisheries.

But under the western cliffs, another ancient occupation, dulse gathering, is still in full swing. In fact, now that dulse has been more widely recognized as a health food—a sought-after commodity in specialty shops on the mainland—the dulse pickers are doing better than ever. Their favourite harvesting location is beneath the cliffs closest to Dark Harbour. It is no wonder that dulse is purportedly imbued with mysterious properties, coming as it does from such a strange and enchanting place.

Dark Harbour could capture the imagination of even the most prosaic individual. For instance, in the mid-nineteenth century even Mr. Perley, a government employee sent to look into the harbour's commercial potential, fell under the spell of this natural wonder. His report is not the usual sort of tedious government document:

> Once within the sea-wall vessels are as completely land-locked, and may ride in as perfect safety, as if in an inland lake, however violently the tempest may rage without; and upon such a precipitous and iron bound shore as the western side of Grand Manan, with nothing but certain destruction to the tempest-tossed mariner who may be cast upon it, this sole place of safety should by all means, and under every consideration of humanity, be rendered easily accessible at all seasons, either by day or by night, and readily found. A few hundred pounds might well be spent in giving perfect access to this most singular and exceedingly safe harbour. The largest ships may lie afloat within a stone's cast of the shore, riding safely with the smallest hawser, while a fearful surf thunders upon the beach without, apparently with sufficient roar and uncontrolled violence, to shake the island to its lowermost foundation. [5]

Because Mananers live in such an awesome natural environment, it is little wonder that they have always been deeply religious. A power beyond the understanding of the mind seems clearly to have been at work here. Yet religious life in the first century of settlement of Grand Manan flowed no more smoothly than the water around the island. Paradoxically, religion caused the islanders considerably more discomfort than did the sea, which they had learned to cope with so readily. At first, they managed well enough with itinerant ministers from the mainland who seemed by and large enthusiastic about their mission to the island and sympathetic to its inhabitants. On Sundays when no qualified preacher turned up, the island schoolmaster read a service. However, increasingly the Mananers yearned for their own church and resident minister. But surprisingly, this yearning produced trouble.

Since the founders of the then-new province had determined that New Brunswick should follow the British pattern of life as closely as possible, adherence to the doctrine of the Church of

The cliffs of Dark Harbour

England was taken for granted. Money was donated for the building of churches and in support of clergymen of this denomination only. Mananers, like other New Brunswickers, were influenced by this fact. Thus, early in the nineteenth century, the Church of England acquired over twelve-hundred acres of woodland on Grand Manan; and the islanders, after raising most of the money for the construction of the church, proceeded to build it with their own hands—once again demonstrating their ability, determination, and self-reliance. By December 1823, the island's first church, Saint Paul's, was ready for its first clergyman: a Mr. Griffin, who turned out to be a disastrous choice.

Mr. Griffin was a quarrelsome Englishman with little love for his fellow man generally, and even less for Canadians—particularly those unfortunate and inept enough, from his point of view, to have gotten stranded on such a God-forsaken rock as Grand Manan. He arrived on the island with his Sisters in the spring of 1824, and for the next four or five years the islanders put up with his nagging, his bad temper, and his lack of understanding or sympathy for their lives and environment. That the islanders' religious zeal survived this man's ministry is, in itself, something of a miracle. Eventually, Mr. Griffin was succeeded by the Reverend John Dunn, a newly ordained clergyman from Saint Andrews who, with his wife and child, took up residence on the island in 1832.

Rev. Dunn was fortunately not at all like his predecessor. He was hard-working, uncomplaining, and faithful. We are told, "in matters spiritual Mr. Dunn achieved remarkable success." But Rev. Dunn's success infuriated the most influential man on the island, Mr. Wilfred Fisher, the magistrate and landowner who had been nicknamed "the Emperor of Grand Manan." Blinded by his jealousy of Rev. Dunn, whom he saw as his rival in wielding power over the islanders, Mr. Fisher took steps to rid the island of the interloper.

On October 9, 1838, an event occurred which would leave the community unsettled for years. On this night the church was burned, and in a tree near it was placed an effigy of Rev. Dunn dressed in "a long black waistcoat, satin vest, a pair of black pantaloons, white gloves and black boots. Arms and legs pinioned by cards to its side, the figure was swinging by the neck…." [6] Everyone recognized the clothes on the effigy as belonging to Mr. Fisher, and so it was upon Mr. Fisher and two of his servants that suspicion for the crime fell. In due course, the three men were tried for their lives at the Supreme Court in Saint Andrews. Despite the fact that their guilt seemed obvious, they were exonerated, much to the disgust of the majority of the islanders. Mr. Fisher even returned to Grand Manan to resume all his former honours, including his duties as magistrate. As the excitement faded, the island's population remained generally demoralized. Cynicism concerning the proper functioning of the law dominated island thinking. Understandably, the years that followed have been regarded as the darkest period Grand Manan has ever known. Although some positive steps were taken, such as the rebuilding of Saint Paul's (in stone this time), it was a time of general despondency and even lawlessness.

In 1844 Mr. Fisher had his way: Rev. Dunn left the island. No one came to take his place, and the bishop seemed reluctant to make the fatiguing and sometimes dangerous crossing to confirm new church members. Interest in the established church flagged. Itinerant Baptist and Wesleyan preachers were received more favourably than before. Gradually, the majority of Grand Manan residents shifted their religious affiliations from Anglican to Baptist. This shift was accelerated by the appearance of the Reverend A. Taylor on Grand Manan. Like an Old Testament prophet, he told of seeing the island in a vision and of being sent there by God:

> Instantly I seemed to be on the deck of a vessel going upon surroundings I had never seen before. "My God," said I, "What is this?" and the answer came immediately. "This is Grand Manan, go there, and you will gather a church of 130 members and see much good done." [7]

Mr. Taylor's vision and ardour proved irresistible. Many Mananers were persuaded that he had indeed been heaven-sent. His inspired and high-powered evangelism left its mark—to this day most of the islanders adhere to the faith into which their ancestors were baptized. Some great spirit must still hover protectively over the island causing it to prosper and yet be spared most of the curses of prosperity. No wonder Mananers are so content with their island. Despite two centuries of settlement, Grand Manan can still inspire in the twenty-first century visitor much of the wonder and delight that the first people sighting this island must have felt.

Notes on Grand Manan

1. "The Perley Report on the Fisheries of Grand Manan" (1850), the *Grand Manan Historian*, no. X, The Grand Manan Historical Society, p. 16.
2. *Ibid.*, p. 20.
3. The *Grand Manan Historian*, no. XIX, p. 25.
4. Of about a dozen small islands in the Grand Manan archipelago, only Whitehead Island is still inhabited. Its population continued to grow because it is on the way to the best fishing grounds. Because of the increase in population, however, the herring gulls no longer nest here.
5. The *Grand Manan Historian*, no. X, p. 22.
6. The *Grand Manan Historian*, no. XXI, p. 26.
7. The *Grand Manan Historian*, no. XIX, p. 34.

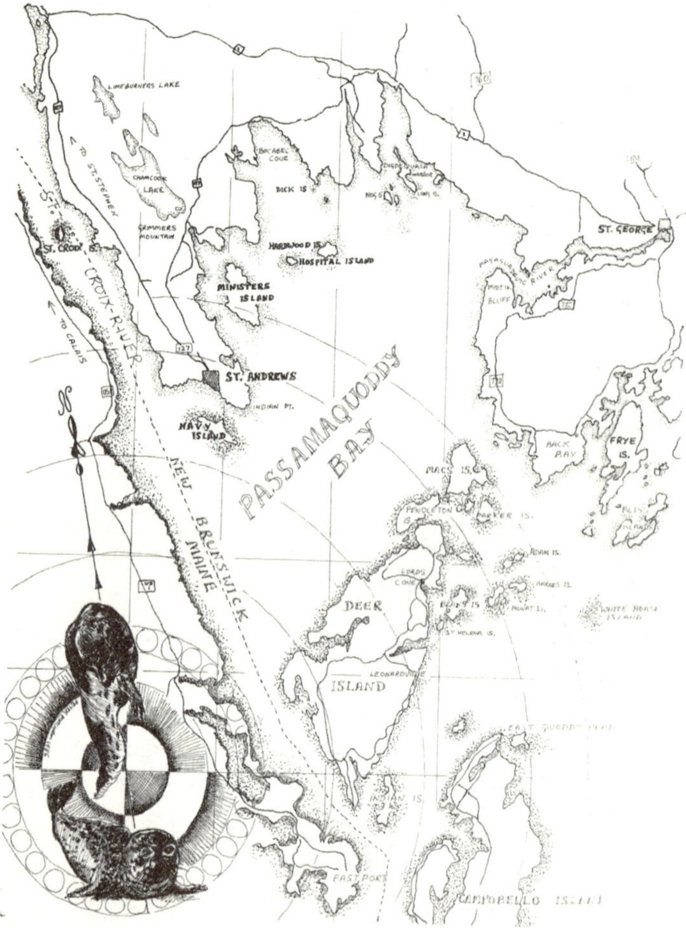

Hospital Island in Passamaquoddy Bay

Passamaquoddy's Hospital Island

We learn that 77 passengers by the Magna Charta *have been discharged from Quarantine Island. 10 deserted, leaving 56 still on the Island of whom 30 are sick, 2 adults and 2 children dead since they landed.*

–St. Andrews Standard (July 28, 1847)

Some ten kilometres northeast of the town of Saint Andrews and well out into Passamaquoddy Bay—three kilometres from Bocabec on the nearest shore—lies a tiny three-acre island, variously known as Little Hardwood, Hospital, or Quarantine Island. But its insignificant size bears no relation to the immense impact it had on the lives of the hundreds of immigrants who were detained there in the mid- to late nineteenth century, when this speck of land served as the quarantine station for newcomers entering the Passamaquoddy region.

Its past is mostly shrouded in mystery, and only odd fragments have survived to hint at the events that occurred there. In books and reports that divulge information about other quarantine islands in the Maritime provinces, no mention is made of Passamaquoddy's

Hospital Island. Even the public archives yield few clues. And yet, here and there in newspapers and in private documents, small bits of information have come to light, giving the bare bones, so to speak, of the strange and tragic happenings on this small and remote island.

Its earliest name, Little Hardwood—one that still survives on some maps—was given because of the island's association with its larger, well-wooded neighbour, which it must have once resembled in miniature. The two Hardwood Islands are only several hundred metres apart, and are connected by a reef some two metres below the surface at low tide. However, Hardwood Island is no longer an appropriate name for either island, especially for the smaller, which is absolutely devoid of trees of any sort. Today not even an alder grows there. Only tall grasses, a few raspberry canes, and sturdy wildflowers cling to the mound that lies above the tide. The larger island, although forested, is now dominated by softwoods—mainly spruce and fir. Here and there birches grow, but the more valuable hardwoods—rock maple, beech, walnut, and oak—are no longer indigenous.

This larger island impresses the visitor as a hospitable spot. With fresh water, a healthy growth of young trees, and still-flourishing lilacs, peonies, and lilies planted by a former owner, it still seems a good place to live. By comparison, Little Hardwood Island, even on a calm, warm, midsummer's day, is bleak and exposed. Although the seals and seabirds, which now make their homes here, seem contented, it is not a place where even the most committed human island-devotee would wish to dwell. Whoever lands here is likely to feel somewhat uneasy, concluding at a glance that few of the essentials for human survival are present. Before setting out to look more closely at this curious mound, he feels that he must ensure that his boat is drawn up high enough on the steep sand beach, and is in no danger of floating off. Such a precaution seems especially well taken when the visitor learns that he has been correct in surmising that this place has no shelter, firewood, or fresh water.

The trees were all cut in 1842 to make way for the quarantine station. The port authorities at Saint Andrews, like their counterparts elsewhere in Atlantic Canada, chose this island site in haste and panic. Here, too, the circumstances were desperate. In February 1842, a vessel had arrived from Europe with sick passengers; they had cholera. Until this time, the Passamaquoddy region, with its salubrious climate and widely dispersed population, had never had to cope with the horrifying epidemics that had ravaged populations in more densely inhabited parts of the world and had left their marks on Canadian cities (like Saint John) and villages (like Chatham). When the diseases, which were current elsewhere, appeared off the port of Saint Andrews, the authorities there were as nonplussed as their counterparts in other localities.

Although by 1842 the practice of quarantining sick immigrants on offshore islands was well established, the Saint Andrews health authorities had selected no permanent site for a quarantine station. When they suddenly had to choose one, they must have assumed that Saint Andrews was going to be better off than the other communities facing the same problem. After all, islands abounded in Passamaquoddy Bay.

Despite the 365 islands reputed to lie just off their town, Saint Andrews authorities soon discovered that not one of these islands seemed right for a quarantine station. The Passamaquoddy Bay residents had always valued their islands, and all of the best ones—those with fresh water, timber, and protected coves with good anchorages, had been granted and settled within a few years of the arrival of the Loyalists. The island lairds, accustomed to pasturing their sheep and cattle on these coastal islands without the expense and bother of fences and hauling in all the fish they required, were not about to be evicted from these precious domains. Moreover, they soon realized that the authorities were not willing to pay a fair price for these valuable properties. Anyway, there was no time to dicker. The newcomers had to be quarantined at once. An uninhabited island—one that no one wanted—had to be found.

It was with this in mind that the Justices of the Peace of Charlotte County fixed upon Little Hardwood; they judged it to be the only island available within reasonable distance of the port of Saint Andrews. Of course, it was not really within reasonable distance of anywhere at all, and partly because of this, Little Hardwood seemed to have had the worst prospect of all the Maritime quarantine islands. Only Partridge Island, just outside Saint John Harbour, had been considered comparable in bleakness, lack of shelter and space, and shortage of fresh water. Yet, even Partridge Island, which Moses Perley, New Brunswick's chief emigration officer during the 1840s and '50s, had pronounced was "an objectionable site," is much closer to land, much more protected, and much larger and more accessible to the port it served than Passamaquoddy's Hospital Island.

One important reason for Saint Andrews' quarantine facility being established so far from the port was that eight years earlier, during a crisis which had fizzled out, the temporary use of an island near town—Saint Andrews Island (now Navy Island)—had sparked a lot of controversy. In September 1834, there had been a local cholera scare. Reports had filtered in that the disease had already appeared in Halifax. Since some ships docked in Halifax before coming to Saint John or Saint Andrews, the Saint Andrews town officials thought they had better be prepared for the arrival of sick passengers in their own port.

The officials held a meeting on September 12 of that year, during which they decided that a health officer with a "good and sufficient boat and two men [were] to be stationed at Saint Andrews Island, and were to remain there until they received further orders." From this station, the officer was to inspect all boats entering the port and examine every passenger. These regulations remained in effect until November 20, 1834, when notices declared that coastal vessels no longer needed to be inspected, and the station was closed down. Nevertheless, throughout the month of September there had been a good deal said and written about cholera: its symptoms, treatment, and cures. One newspaper advertisement read:

> Cholera medicine of tried and known efficacy in small packages for families put up at the shortest notice.
> –W. C. McStay, St. Andrews, 23 September, 1834

No one seemed to object to such bogus advertising. In subsequent issues of the paper, no explanation was offered to inform readers that claims of this sort were misleading and that no reliable drug existed for curing cholera. Meanwhile, by September 25, the local board of health had issued elaborate warnings about the symptoms of this disease and the need for quick action in combating them. Failure to take care, they bugled, was usually fatal.

Such frightening publicity obviously did nothing to popularize the building of a permanent quarantine station so close to town as Saint Andrews (Navy) Island. Besides, the island had already been granted, and its owners were disconcerted with having their territory encroached on. The views of the two major proprietors—the military and the church—carried a lot of clout. Both groups argued that the 140-acre island was too large and valuable a property to be given over to an immigrant detention centre. It was easy for the majority to support such claims when everyone was worried about having the quarantine facilities almost at their doorsteps. Town residents had already heard tales of passengers escaping from other hospital islands, and were well aware of the threat that such potential runaways posed.

Once the quarantine station was established away from the immediate vicinity of the most important Passamaquoddy communities, and the immigrants banished to the location often vaguely referred to in reports as the "lower bay," public attention turned elsewhere. The focus of newspapers shifted back to world events, local comings and goings, and occurrences in other parts of the Atlantic provinces and their neighbouring New England states. Some reporters drew attention to the large numbers of immigrants expected throughout North America due to the economic crisis in Ireland; others catalogued the tragedies and traumas taking place at quarantine stations on Grosse Île (Quebec) and on the Miramichi islands; and still others recorded the cholera outbreaks in Southern Russia, where epidemics had wiped out entire villages. Yet, about local quarantine conditions, there was hardly a word. Even books published especially for emigrants—notably Abraham Gesner's *New Brunswick with Notes for Emigrants* (1847)—fail to mention Passamaquoddy's Quarantine Island. Gesner's description of the bay focuses entirely on its idyllic aspects:

> The whole of Passamaquoddy Bay is studded with islands, which are said to be 365 in number. Many of these islands are merely rude masses of rock, or small eminences covered with moss and stunted spruce. The larger ones have a soil of medium quality, and produce excellent potatoes, barley and oats. During the summer or fishing season, the bay presents an interesting spectacle. Boats and vessels becalmed are swept away by the rapid tide. At one instant they are hidden by some blackened rock, at another they are seen gliding from behind the green foliage of some little island. Some hundreds of boats are huddled together, their crews being actively engaged in drawing forth the finny inhabitants of the deep. As soon as the shoal of fish retreats, or the tide is unfavorable, they disperse, and the surface of the water is decorated with their white and red sails. The Indian, in his frail bark canoe, without a rope or anchor, is also there, and the report of his gun, discharged at the rising porpoise, is re-echoed among the cliffs of the shore. Flights of gulls hang over the glassy surface of the water, which is here and there broken by a shoal of herring, or the spouting grampus in search of his prey.

Throughout the disastrous year of 1847, there was only one brief mention—in the *St. Andrews Standard*—of the local Quarantine Island: "We learn that 77 of the passengers by the *Magna Charta* have been discharged from Quarantine Island. 10 deserted, leaving 56 still on the Island of whom 30 are sick, 2 adults and 2 children dead since they landed" (July 28, 1847). But apparently no reporter was sufficiently intrigued by the discrepancy between the auspicious name of the ship and the inauspicious destiny of so many of its passengers to bother investigating. Nothing more would appear in connection with the fate of these people or of passengers from other ships except for one other short entry on August 11 of the same year, which also pertained to the *Magna Charta*: "The carpenter of the *Magna Charta* died at the Marine Hospital on Thursday last aged 55." No clues remain as to how ten of the *Magna Charta*'s passengers managed their escape, or as to whether or not they survived. Did they take the doctor's boat? Did they try to swim? Despite the midsummer weather, their chances of reaching the mainland were poor. The cold water, high tides, and frequent fogs were all against them. Oddly enough, a successful escape from this island would have been easier in mid-winter when, according to local tradition, the ice between Hardwood Island and the mainland was sometimes strong enough to support a herd of cattle driven across it. Since ten passengers from one ship defected at once, it would seem that few security measures were enforced off the island—no armed patrols comparable to those off other quarantine stations. But then, one look at the island's position would have been enough to convince even the most suspicious person that such measures were unnecessary.

In 1848 there was nearly no news of Saint Andrews' Quarantine Island, despite the fact that the barque *Star* had landed a large number of sick passengers there that year. Although forty-eight Irish immigrants had died and been buried on the island, a public report of the quarantining of the *Star*'s sick passengers was delayed until late August 1903—that is, some fifty-five years later. Then indeed, the *Saint John Sun* gave it considerable coverage. The *Sun*'s early-twentieth-century readers were told that the Earl of Fitzwilliam, County Wicklow, Ireland, had sponsored the passengers on this ill-fated ship. Fitzwilliam seems to have been much more benevolent than many of his peers, since he had not only paid for these immigrants' passage, but had also set aside funds to maintain them during the first six months of their residence in New Brunswick. Those who survived the fever, we are told, were "a hardy, long-lived lot of people," despite the slavish labour that the men expended on the construction of the New Brunswick and Canada Railway (in which Fitzwilliam was a shareholder) for some time after their arrival. The early-twentieth-century chronicler of these events called attention to the troublesome fact that no monument had yet been erected to mark the resting place of the immigrants who had perished on Hospital Island and pointed out that such a marker had long since been placed on Quebec's former quarantine island, Grosse Île.

From its inception in 1865, Saint Stephen's newspaper the *Saint Croix Courier*, was just as remiss as the *St. Andrews Standard* had been about reporting on occurrences at Quarantine Island. For instance, in the *Courier* of 1865, the year in which Hospital Island became recognized as the quarantine centre for the whole of Passamaquoddy Bay, almost every other subject—recipes, imported commodities, American and European news, gossip, Fenian raids—took precedence over the local quarantining of ships and outbreaks of communicable diseases aboard vessels entering Passamaquoddy Bay. Two terse sentences in one issue of the *Courier* (1865) were devoted to the quarantining of the SS *Atlantic*:

> Dr. Burdett, Quarantine Surgeon, reports that 15 deaths by cholera out of 50 or 60 cases on board the steamship *Atlantic*, occurred on the passage from Brest. The *Atlantic* has been sent to the lower bay, and all communication with her forbidden.

Yet, in an issue of the same newspaper the following year, a whole paragraph is devoted to the cholera outbreak on board the *England* in Halifax Harbour and the measures adopted to deal with it. Meanwhile, newspapers such as Saint John's *New Brunswick Courier* and Chatham's *Gleaner* regularly printed news of the quarantine island closest their communities, perhaps because these islands were so immediately visible to anyone coming or going through these ports that any unusual occurrences simply could not be overlooked.

Nevertheless, given the concern about immigration and the spread of communicable diseases manifest in the *St. Andrews Standard* throughout the 1830s, it seems surprising that when, on March 10, 1840, forty acres on Hardwood Island were put up for sale, no one suggested buying them for a quarantine station. The larger island, with its cleared meadows ("cuts 15 tons hay annually," read an advertisement in the *Standard*) was obviously a much more suitable choice than its smaller sister.

Despite all of the problems establishing a quarantine station on Passamaquoddy's Hospital Island, the buildings constructed there seem to have been a good deal more elaborate and substantial for their time than those used for the same purpose on other Maritime islands during the mid-nineteenth century. The first hospital, built in 1842, was fifteen by eight metres large, and it served the region until 1848 when a second, more adequate building was added. This new structure was begun after the desperate season of 1847, when quarantine facilities everywhere on the Atlantic seaboard—as far apart as Grosse Île, Middle Island, and Ward's Island (New York)—were so overtaxed with successive waves of sick immigrants that authorities in all locations were made to face up to the pressing need for improved conditions, both for the newcomers and for those who cared for them during their incarceration. In all of the detention centres, 1847 had been a record year for deaths of quarantine doctors as well as their charges. Clearly, something had to be done. Just down the Fundy coast from Saint Andrews, Moses Perley and three or four Saint John physicians were planning and agitating for improvements on Partridge Island. Change was in the air.

The new hospital on Quarantine Island was exceptional for its time and place, contrasting sharply with the wretched sheds used on Middle and Sheldrake Islands at the time. Its specifications closely resembled those laid down by Perley and the Saint John medical establishment for a new quarantine centre on Partridge Island. Obviously, the Saint Andrews Hospital Island facilities were on a smaller scale than those on Partridge Island, since immigration through the port of Saint Andrews never equalled that which occurred through Saint John. Yet, the new hospital on the Passamaquoddy Island was a two-storey structure, eighteen by eight metres, with a basement, plastered walls, and two chimneys. In keeping with Perley's recommendations at this time for Partridge Island, the Hospital Island facility also had a doctor's house, a keeper's residence, a shed, and a sailboat. From 1848 on, the original hospital was supposed to be used only as a summer hospital.

Putting out from Holt's Point, Bocabec, for Hospital and Hardwood Islands in a six-metre wooden whaler, the contemporary individual would have little difficulty imagining himself back in the early nineteenth century when the larger island had a single owner, and its smaller, sister island was uninhabited. Mr. Holt, the nearest mainlander and a descendant of the settler who gave his name to the point, once kept an eye on the residents of the larger island on his way to the fishing grounds. This was a sensible and neighbourly precaution in view of the life-threatening dangers of fires, accidents, smugglers, and pirates to which dwellers on these remote isles have always been vulnerable. In the cove where former Holts moored their boats, their late twentieth century descendant kept his, hauling it in to shore hand-over-hand past the weirs using the same time-tried pulley arrangement as his ancestors. So although the original white pines no longer stand along this point and a small, eighteen-horsepower engine powered the boat, the illusion of embarking on a journey back in time persisted.

Mr. Holt, though of retirement age, walked sure-footedly into the water over submerged rocks so slippery that his two passengers, picking their way, not only fell behind, but could scarcely remain upright. He placed a large, boxed compass and two life jackets on the bottom boards of the open boat, climbed in, and patiently awaited his companions who, though not many metres away, remained indistinct, ghostly shapes hovering in the dense fog. He smiled a little when the first guest to climb into the boat asked why he carried only two life jackets. He explained that he saw very little point in taking enough preservers since they merely prolong the misery of drowning in these icy waters. In the silence that ensued following this sobering explanation, Mr. Holt started up the motor and steered away from the weirs and into the fog. Once, he nodded towards a denser, darker shadow: Hog Island.

The motor hummed, seeming louder than normal as the reverberations bounced back off the walls of fog. Then Mr. Holt cut the motor. The absolute silence, the dead calm, the encircling walls of fog—all seemed ominous. Then, out of the fog and very close by, loomed land, low and rocky. As the strangers strained to see through the heavy curtain of fog, it parted enough to disclose perhaps thirty or forty blackened and lifeless tree trunks, still standing as if in the aftermath of a fire. Although the sight of these dead trees festooned in streamers of fog was not so remarkable in itself, what transformed the scene, lending it an eerie fascination, were the hundreds of large, long-necked, black seabirds—strange, prehistoric-looking creatures—perched on the upper branches of the trees, while hundreds more swirled above them.

"Dick Island.... Shags nesting," Mr. Holt announced both proudly and cryptically, as if unveiling a priceless work of art for friends to admire. "You might call them cormorants."

Soon after leaving the cormorants' island, Mr. Holt beached his boat on a steep slope of sand and gravel: Hospital Island. He pointed out the creatures he knew: the great black-backed gulls and the mackerel gulls they bully; the cranes (herons) poised on the beach; and the seals bobbing in the water. He called attention to the fact that the island is smaller than it was before World War II because "the army blew the end off it in target practice."

Hospital Island does seem pathetically small and vulnerable—to rain and sun, and to every storm sweeping up the bay. A modern mortal standing atop the grassy knoll shrouded in fog may imagine himself surrounded by the ghosts of those who died here long ago, too weak to cross the threshold into a new life. This is perhaps as close as he can come, in a different time and position, to glimpsing the loneliness, the dreams, and the nightmares of the straggling procession of vulnerable humanity who were thrown up upon this mound. The more he broods upon the past, the more clearly he understands that the only creatures fitted for a contented survival here are the present inhabitants: the nesting gulls, the herons fishing along the shore, and the seals who frolic among the rocks.

Cormorants in the Bay of Fundy

Partridge Island

I perfectly concur in the opinion of the Medical Gentlemen, that Partridge Island is an objectionable location for the Quarantine establishment. The island is but small, and the most eligible portions are occupied by the Light House, the Keeper's residence and grounds and the Military Works. There is very scanty supply of water upon it, and it is doubtful if Wells would furnish a sufficient supply of good fresh Water.

Partridge Island is difficult to access at all times, except by day light, and in fine summer weather; in stormy weather it is positively dangerous to visit it. The expense of sending supplies of fuel, straw, fresh water, provisions and stores to the Island, has been very considerable during the past season, while delays frequently occurred during stormy gales, or when a heavy sea was rolling from the Bay of Fundy, or when a dense fog prevailed.

–Moses Perley, Provincial Emigrant Agent, Port of Saint John (1848)

The Island [Partridge]...is undoubtedly one of the best sea resorts in the maritime provinces. It possesses many advantages which the usual resorts of tourists are deficient of. The walks about its woods and shores are delightful, and almost any portion of the land commands a splendid view of the bay or the city....

–THE DAILY TELEGRAPH, SAINT JOHN (1894)

We were placed on Partridge Island in barracks and I think it was one of the most dismal places to keep a crowd of men.

–PHILLIP McBRIDE, GUNNER, FROM HIS DIARY (1915)

It's a lovely place. I'd hate to leave it [Partridge Island]—hate to be anywhere else.

–MRS. HAROLD LAUDER, WIFE OF THE LIGHTKEEPER AND ASSISTANT KEEPER HERSELF DURING WORLD WAR II (1948)

Is it possible that these people were all discussing the same place? How could four reasonable human beings, over the course of a century, support two such diametrically opposed views on one small (twenty-four-acre) island? The reason for these very different assessments of Partridge Island—as an ideal resort or as a prison—has to do with the surprising diversity of individual experiences there; some have clearly been of the most pleasant sort, and others disagreeable in the extreme. The emigrant agent, newspaper reporter, soldier, and lightkeeper have put forward views representative of the time in which each lived and of the roles each filled in society.

Because Partridge Island has been most publicized as a quarantine station and a military stronghold, it is the aesthetically unpleasing images of this island that have stuck in the minds of the majority of those who have thought about this rocky, treed oval that the legendary Glooscap is supposed to have wrenched up and tossed down from the river's mouth after it broke up the beavers' dam there so long ago. The confinement of large numbers of people in hospitals or barracks is never an attractive prospect. Add to this the problems that Moses Perley focused on—a lack of fresh water and a shortage of space for facilities expansion—and one is face-to-face with Lescarbot's vision of islands as prisons.

Publicized escapes from the island detention centre reinforced this unappealing image:

> We trust the authorities will insist upon a rigid observance of the quarantine regulations, and severely punish all transgressors. We understand that John Lawson and James Cullan, seamen, and John Burns, a passenger, made their escape from the barque *Amazon*, at quarantine, last Saturday night. Means should be taken for apprehending.... (*New Brunswick Courier*, June 5, 1847)

Reports such as these make comparisons between incarceration for carriers of communicable diseases and imprisonment for criminal offenders inevitable. In 1848, when Perley complained about Partridge Island's shortcomings as a site for a quarantine station, he was at his wits' end. By then he had been chief emigrant agent for New Brunswick for five years, and in that time he had heard more tragic tales relayed to him from around the province than he cared to remember. He had also seen a good many tragedies enacted before his eyes.

Moses Perley saw the Saint John quarantine facility as his most pressing responsibility. A dynamic, kind, and principled man with a utopian dream for his native province and its inhabitants, Perley felt compelled to improve conditions on Partridge Island and

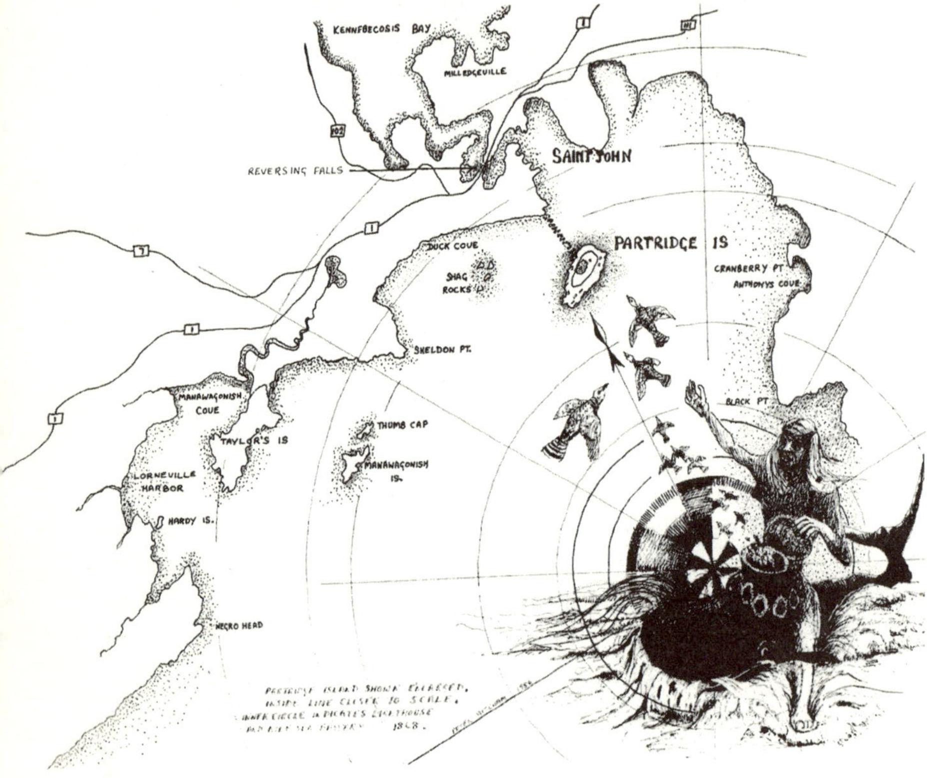

to find suitable land for newcomers who had neither friends nor a chosen destination. If he failed to help, he knew that the immigrants who survived their ordeals at this station would never be able to share his vision of New Brunswick as a potential Eden. Anyway, as a humanitarian, he was not prepared to witness another year as disastrous as 1847.

The Partridge Island Quarantine Station was the facility through which the majority of newcomers entered New Brunswick. In 1847 alone, more than sixteen thousand immigrants, mostly Irish, arrived here—a huge increase over the approximately two thousand new arrivals at this port facility in 1844; the six thousand in 1845; and the nine thousand in 1846. What seems remarkable about the whole situation, judged from a contemporary standpoint, is that the authorities managed to cope with such a sudden influx of people. How did they ever make do with the twenty-four by six-metre "pest house," built after the 1830 smallpox epidemic, and a second hospital, constructed subsequently for cholera victims? These two facilities accommodated no more than two hundred people altogether, and by June 21, 1847, there were 2,500 immigrants being held in quarantine on the island. The army provided tents for some of those left outside, but many of the newcomers were too weak to put them up. Sick and starving, they lay on the bare earth exposed to wind, sun, and rain.

July and August 1847 were the worst months for severe illnesses and fatalities. Not even the doctors were spared. Young Dr. Collins died on July 3, and for much of August both Drs. Harding were too sick to carry on working. Not only were there not enough doctors and nurses to cope with the sick but, for a while, there was no one to bury the dead. On one day, forty corpses were piled up in the "dead house" awaiting coffins and burial. The corpses were never put into coffins, but stashed in a long trench that served as a communal grave; they were placed there without ceremony, with earth simply shoveled over them. It is little wonder that Moses Perley's heart sank when he pondered such appalling facts and looked back over accumulated notices attesting to even more widespread suffering:

> CONVICTION UNDER THE PASSENGERS' ACT (The Queen vs Austen York) On Saturday last the defendant in this case, Master of the barque *Linden*, from Galway, was convicted...in not having issued water and provisions to the Passengers on board his ship.... (July 3, 1847)

> Reports of impositions upon the immigrants by the boatman in the employ of the Corporation have been circulated.... (July 3, 1847)

> As Small Pox of a very virulent character is raging at the Quarantine Station, it is to be hoped that the advantage will be taken by our citizens of the benefits of vaccination.... (August 14, 1847)

> The Visiting Physician at Partridge Island reports that there are now 800 emigrants upon the Island of whom 600 are sick.... (Government Emigration Office Saint John August 1847)

Depressing as all the publicity was, it kept the problems relating to the quarantine station almost constantly before the public and the politicians. As a result, the citizens of Saint John continued to demand solutions to the problems they had been made aware of. They even took a hand themselves in raising money for immigrant relief. Clubs and societies sponsored charity balls, concerts, and bazaars.

From a contemporary viewpoint, such philanthropic enterprises are not without their ironies. For instance, while one of these balls—"by far the most brilliant affair of the season"—was in progress, would-be recipients of the money raised were dying aboard immigrant ships for lack of life's necessities. Reporters who enthused over the four hundred ballroom guests, amongst whom they noted "so many bright eyes, joyous faces and well-dressed ladies," were soon to record the arrival of the barque *Alderbaran* at Partridge Island, the death of thirty-four of the children aboard at sea, and the horrific prospect of "upwards of 100 [out of 418]...ill with fever and dysentery."

The reporter who was entranced by the bright eyes and fine clothes at the fashionable ball was not the only responsible citizen whose focus in 1847 was not on the quarantine station and the suffering immigrants incarcerated there. In that same year, Abraham Gesner, a medical doctor and an acquaintance of Moses Perley, published a commentary on Partridge Island and its environs without focusing on the immigrants, the limited facilities in which they were

Glooscap breaking the beaver's dam and creating the Reversing Falls

lodged, or the diseases from which they suffered. He did indeed note in passing that a quarantine station was located on the island, but nothing more. His main focus was on the valuable fishery surrounding the island. Here, he noted, cod, halibut, and pollock could be caught all year, and shad, alewives, and salmon in spring and summer. Dr. Gesner even speculated about how the island must, at one time, have been joined to the nearby mainland: "Between the island and the mainland westward there is a long narrow bar, dry at low water." A contemporary reader cannot help feeling that this is a surprising commentary from a medical man in a plague year—especially in a book aimed at immigrants (*New Brunswick with Notes for Emigrants*).[1]

Such incongruities are, of course, always easier to spot at a distance—and the fact is that a great many Saint John residents did expend considerable thought and effort on the newcomers entering their city. When, later in 1847, a boatload of children in the care of old women was put ashore at Partridge Island and later removed to the poorhouse, hundreds of Saint John women, many of them in difficult circumstances themselves, set about making clothes for the destitute youngsters. Others, like Moses Perley and his wife, took abandoned children into their own homes despite the very real risks of infecting themselves and their own children. (Perley did in fact contract the fever and several of his children died during epidemics that had spread from the island to the city.)

In view of all the heartbreak surrounding the plight of the young waifs sent out to a new land without protection or worldly goods, Perley's bitterness on this subject is understandable. His state of mind is reflected in characteristically brusque statements: "One hundred and twenty orphan children were thrown upon me yesterday from the *Lady Sale* about naked. I have taken 2 boys and 2 girls—who wants orphans?" So volatile did the emigrant situation become in 1847 that the Lieutenant-Governor himself agreed to look at the Partridge Island facility and, if necessary, recommend changes. The mayor and aldermen were anxious to have the townspeople told that they had been doing their duty: "...supplies for the sick have been promptly supplied by the proper authorities and too much praise cannot be awarded to His Worship the Mayor and Mr. Alderman Smith for their active personal exertion in the cause of suffering humanity..."

Because of the general concern of the citizenry, stirred up by the press and by those in charge of the quarantine station—particularly Moses Perley and the doctors—Partridge Island was in the process of becoming the best-documented of all the mid-nineteenth century quarantine facilities along the Atlantic coast, as well as the safest. Numerous as the deaths were there, they were, by 1853, just a fraction of those at the many other quarantine stations, including the facility at Ward's Island, New York. This fact is particularly surprising in light of the evidence that sick immigrants were turned away from this and other American immigrant processing centres and redirected to Saint John, where they were accepted.

In his 1853 year-end report, Perley expressed his satisfaction with the improved physical conditions at Partridge Island. He also summarized the benefits passed on to the emigrants entering through this station because of the legal battles he had won over the preceding decade against ships' captains trying to take advantage of helpless newcomers. Since, under the Passengers Act, he had obtained convictions "in every case of willful violation," Perley noted that: "prosecutions are rarely necessary, full compensation being generally obtained by my interference or request."

The comparisons Perley drew between Ward's Island and Partridge Island, backed up as they were by firm figures, are surprising. First of all, he pointed out that emigrants entering the United States who had been robbed and mistreated on the voyage, "have no remedy at New York, the shipping interest being too powerful.... Not a single prosecution took place there last season, although the most flagrant cases had excited much attention from the public and the press...." Moreover, on New York's quarantine island, Perley emphasized, the death rate was "no less than seven

The quarantine hospital at Partridge Island

percent on the numbers embarked;" whereas, for emigrants entering New Brunswick, "the whole number of deaths on the voyage and in Quarantine was less than one third of one percent on the numbers embarked. The deaths were chiefly among infants and adults affected with chronic complaints of long standing." What Perley did not say much about, and what obviously contributed to the difference in mortality statistics, was that he had personally helped hundreds of immigrants by giving them money and clothing and by settling them on land that he had himself surveyed. Although the government was supposed to reimburse Perley for such aid, it rarely did.

Yet Moses Perley's enthusiastic assertion at the end of 1853—that the most serious problems connected with the Partridge Island quarantine station had been brought under control—was premature. In making his year-end statement, he was not to know that he stood on the threshold of a plague year that was to rival 1847. Only the disease—cholera—was different.

Immigrant ships standing off the quarantine facilities on Partridge Island

By April 1854, when the *Blanche* arrived from Liverpool with a shipload of German immigrants, fifty passengers had already died. Until June the disease was contained on Partridge Island and on the ship. Then, as was to happen in Halifax just over a decade later as the result of cholera aboard the *England,* infected clothing floated in to the mainland and was claimed by impoverished city dwellers. However, whereas in Halifax the disease would be contained by Dr. Tupper's swift action in isolating the mainland victims, in Saint John it spread rapidly through the population. In New Brunswick's great port city too many people had been exposed simultaneously at the beginning.

Among the first victims Dr. Harding diagnosed were a local policeman and a sailor from the *Blanche.* The doctors and their assistants were then kept busy rushing from house to house trying to control the pestilence, but they were not very successful in stemming its course. The epidemic had got a head start. Throughout July, August, and September, cholera claimed more than 1,200 victims, some of them survivors of early fever ships.

Because of this cholera outbreak, regulations governing ships entering quarantine off Partridge Island became more stringent than before and, as a consequence, contagious diseases were henceforth mostly confined to the island station. Even there, until the centre closed in 1938, there was usually only a handful of sick people suffering from assorted illnesses: scarlet fever, smallpox, measles, and, in 1902, yellow fever. The only notable exception to this pattern occurred at the turn of the century (1900), when the quarantine station was overburdened with smallpox patients and a new hospital had to be built to house them. Then, briefly, the doctor and the nurses who were called in worked night and day caring for the sick.

Peaceful Interludes

However, both before and after this outbreak, life for the island's full-time residents progressed at a tranquil pace. During these peaceful interludes, the lightkeeper resumed his customary role as the island's chief potentate. Like his predecessors—the chosen few who had regulated the light since it was first lit in 1791—James Wilson presided over his domain wisely and benignly.

Mr. Wilson had arrived on Partridge Island in an immigrant ship toward the end of the 1854 cholera epidemic. A practical, good-natured twenty-two year old from Glasgow, he had come out to New Brunswick "to fill an engagement with the gas company." However, he never did get beyond the island; for when government officials learned that he was used to working with gas, they managed to persuade him to take charge of the gas house, which supplied the light on Partridge Island. He proved to be so competent that he eventually replaced Mr. Reed, who had been lightkeeper upon his arrival.

Throughout his long tenure as lightkeeper, James Wilson demonstrated outstanding competence in the important post that he had been asked to fill. He never forgot for an instant that thousands of lives depended on the care and skill that he put into the mechanisms he was responsible for. As a 1894 reporter noted admiringly: "The lighthouse is...always kept in splendid order. No light on the coast can be seen farther from the sea, and since Mr. Wilson's stay on the island it has never been out from sunset until sunrise."

Keeping the light burning steadily was perhaps not Mr. Wilson's most remarkable achievement, particularly once paraffin had replaced gas in the burners.[2] The trickiest part of his work seems to have been keeping the temperamental fog alarm going. As a journalist observed: "The fog alarm building contains two boilers, one of which is so unsafe that it cannot be used. With the other boiler, however, Mr. Wilson and his son, who now acts as his assistant, manage to keep up enough steam to blow the whistle, which in foggy weather blows 10 seconds of every minute." However, perhaps the most ingenious of all the devices that James Wilson introduced over the years (and certainly the one that most interested the visiting reporter of 1894) was the small wooden railway he had constructed to convey supplies from the cove on the east side of the island to the lighthouse and fog whistle. For years, until the government built the wharf facing the harbour, he had landed all supplies in the cove and moved them along this track.

But by June of 1894 the government wharf had long since become the established landing place. Alongside it was moored the doctor's boat, the *Neptune*, which was used for boarding all incoming vessels. At this time the Wilsons shared Partridge Island with only one other couple: Thomas and Margaret McGowan, and several of their six children. Mr. McGowan, five years older than Mr. Wilson and related to him by marriage, had come to the island in 1878 as assistant lightkeeper. Eventually, when Wilson's son was ready to help his father on the light station, McGowan took the job of quarantine steward.

In June 1894, the Wilsons and McGowans were looking forward to sharing the island with young Dr. March, who was about to take up Dr. Harding's duties as quarantine physician. He was, a reporter noted, to occupy "the very pretty and pleasantly situated cottage on the island." Dr. March did not expect to have to cope with the same sort of nightmarish conditions that both Drs. Harding had endured while they had been resident quarantine doctors. It had been some time since one of the old hospitals had burned and another had been dismantled, and there was no talk of replacing either of them. One of the original hospitals still stood, as well as the old marine hospital (which resembled a small house), a disinfecting shed, and an oblong building—which in an emergency could be used as a hospital—but no one really expected that any of these buildings would ever again be filled to capacity.

Indeed, if the reporter who visited Partridge Island in June 1894 is to be believed, most Saint John residents at this time tended to regard their offshore island primarily as the idyllic destination of "numerous picnic parties." Few people wanted to recall the bad old days of the thirties, forties, and early fifties. Certainly the townspeople's' revised assessment of the island seemed justified, for despite the fact that Dr. March inspected more than 74,000 immigrants and ships' crews in 1894, only a handful of the new arrivals were placed in quarantine. During his first few years on the station, the doctor generally spent more time entertaining ships' officers on the island, or being entertained by them on board their ships, than he did looking after sick newcomers. This did not mean that he was remiss, but simply that times had changed.

During the next few years, the majority of the immigrants entering Canada through the port of Saint John passed Dr. March's inspections aboard ship without ever setting foot on Partridge Island. Among the most talked about of those who were examined while their ship rode at anchor off the island were the two thousand Doukhobors who, in January 1899, had been quarantined on Lawlor's Island in Halifax Harbour.

The Doukhobors' coming had been much publicized in the Saint John newspapers early in February, both because of their connection with Count Leo Tolstoy and because of their unusual philosophical and religious beliefs. Saint John residents regarded these newcomers with a mixture of curiosity, benevolence, and intolerance—feelings much akin to those that were demonstrated on the arrival of large contingents of Irish immigrants in Saint John more than half a century before. People persecuted in their homeland and obliged to forsake it and take passage for a strange new country in the dead of winter are almost bound to arouse a variety of emotions in those who have the time and inclination to sit back and ponder such a situation.

Although this contingent of Doukhobors (the second to enter Saint John) had experienced a quick and almost uneventful crossing from Russia (marred by a single smallpox death), followed by an un-traumatic sojourn at Halifax's quarantine island, Saint Johners were eager to learn how these people had endured the rigours of their voyage from Halifax to Partridge Island. Day after day the residents of New Brunswick's great port city read the shipping reports in which the *Lake Superior* was listed as still en route, without its position pinpointed. While they waited, the townsfolk were bombarded with startling reports of particularly wretched weather on land and at sea and the onslaught of "la grippe." The headlines were all bad: "Record Breaking Cold...," "Ice on Ships Entering Port...," "Steamers, Barques and Schooners Ashore on the Rocks...," "A Fearful Storm...." Finally, on February 18, the *Saint John Globe* published the long-awaited bulletin:

> S. S. *Lake Superior* passed Briar [*sic*] Island at 9:30 o'clock this morning. At 3 o'clock this afternoon she was in sight of Partridge Island and was expected to reach the quarantine station about 3:30 o'clock. She will there be boarded by Dr. March, port physician, in the quarantine tug *Neptune*...preparations at Sand Point to give the 2,000 Doukhobors quick dispatch.

The *Lake Superior* had taken twenty-three days to steam from Halifax to Saint John, the same length of time as the voyage from

Russia to Halifax. Because of the gales, blizzards, and extreme cold, the journey had been as arduous and dangerous as any in the dark old days of the early part of the century. The curious, the benevolent, and the critical went out to watch the Doukhobors disembark. How, everyone wanted to know, had they withstood the ordeal? Well…very well, by all appearances. The adults who jammed the decks were singing hymns, and some observers who had read reports of the Doukhobors' departure from Halifax must have wondered if these hardy immigrants had stood on deck singing for the entire voyage. Not only was the morale and health of these immigrants visibly superior to that of their mid-century Irish predecessors, but their orderly disembarkation at Saint John in no way resembled the haphazard spilling out of the earlier Celtic migrants. The welcome was also different—and better—for the turn-of-the-century immigrants.

At the dockside, representatives of the local council of women stood giving out treats—bags containing an orange, three or four "substantial cookies, and a few raisins and candies"—to the Doukhobor children. Initially the ladies had made up seven hundred bags, but these were dispensed so quickly that local dignitaries had to scurry around to find additional goodies for the crowd of children unsupplied—an indication of just how many of the two thousand newcomers were children. Journalists noted the "pleasant looks on the faces of the mothers who watched their children treated," and reported that "the ships' company all speak in high terms of the Doukhobors."

After the treats, the whole group passed directly into the trains waiting to take them to their western destination. The commissioner of emigration had even come all the way from Ottawa to supervise the arrangements. Yet despite the efficiency and kindness of certain officials and welcoming groups, not everyone felt warmly about the Doukhobors. Perhaps because the newcomers did not speak English and their beliefs seemed so strange, some dockside observers viewed them as a species apart from themselves. This attitude was reflected in several newspaper accounts of the time, which recorded the Doukhobors' arrivals and departures as if these idealistic immigrants had been just so many packages processed for shipment:

> Dr. Montzambert (chief national quarantine officer) will then have them loaded on the steamer again (in Halifax) and brought to Sand Point (Saint John) where they will be forwarded to their destination as were the first lot....
> –The *Saint John Globe*, January 10, 1899.

> Mr. Frank Pedley, Commissioner for Emigration, who arrived on Thursday from Ottawa has completed all arrangements for the transshipment for the second lot of Doukhobors....
> –The *Saint John Globe*, February 17, 1899.

The Military Occupation

As the early years of the twentieth century slid by with ever fewer crises and casualties at the quarantine station, the island's permanent residents seemed justified in believing that nothing much would ever again disturb their peaceful existence, save the blast of the foghorn warning mariners to steer clear of the rocks. But they were wrong. New European upheavals—two World Wars—were in the making; they would send shock waves as far as this tiny island and change its face forever.

With the increasingly sophisticated weaponry of the twentieth century and the advent of aircraft and submarines, distances between continents seemed to be shrinking. Even a port as remote from European theatres of war as Saint John, New Brunswick, might, some strategists argued, be vulnerable to enemy incursions. Deployment of significant numbers of troops on the island outside the harbour was considered a sensible precaution.

Such reasoning was not exactly new. Provision for a military battery and garrison on Partridge Island had been made soon after the arrival of the Loyalists and in former times of crises. In 1800, 1812, and intermittently afterwards, there had been a flurry of military activity there. On these occasions, several officers and a handful of men had been stationed on the island to keep watch for possible invaders and preside over gun emplacements. But with major conflicts in the offing, correspondingly large troop deployments seemed necessary—even in such an out-of-the-way place as Partridge Island.

Although the island was small, it had, the authorities pointed out, facilities for housing large numbers of people. The quarantine hospitals could be converted readily into military barracks. This was done promptly, and soldiers (at one time as many as fifteen hundred) were soon lodged in the old buildings where immigrants had languished not so many years before. Just as the ghosts of these former tenants seemed to haunt the buildings' interiors intermittently, tangible reminders of their collective fate were discovered outside with startling regularity. As the soldiers dug through and blasted aside obstructions to make way for new fortifications, they regularly came across the bones of immigrants who had perished during their island incarceration. Skulls, thigh bones, tibias, and fibulas turned up daily as reminders of these past tragedies and of the ever-present vulnerability of humanity.

If such discoveries did little for the morale of the troops, the confinement of so many people in so small a space was equally demoralizing. Phillip McBride, in his World War I diary, described the irksomeness of the soldiers' daily routine on Partridge Island:

> On fine days we marched up and down by the hours and drilled with rifles and bayonets and on wet days listened to hours of tiresome lectures which never did us any good as they were all about the science of gunnery which was useless to any except the officers, or about knotting and lashing ropes which was 20 years behind the times.

If many of the soldiers were depressed by their stay on Partridge Island, the Lauders, lightkeepers of the island during World War II, grew to enjoy the activities the military sponsored. As Mrs. Lauder recalled, "We had no trouble with them [the soldiers] at all. And Partridge Island was really lively—a picture show or some other kind of entertainment every night."

So it has been through the years: every negative view of Partridge Island has had its positive counterpart. Most of the year-round residents; their guests; and chance picnickers who spent fine summer Sundays exploring the coves, watching ships entering or leaving the harbour, and searching for treasure, loved this place. Others, part of large contingents quartered in barracks or hospitals, forbidden to leave, and generally deprived of the simple delights of more tranquil times, hated it.

The large groups of soldiers and immigrants who sometimes thought themselves no better off there than prisoners, had to contend periodically with the same sorts of difficulties that had plagued Champlain's men on Dochet Island. Paramount among these was a shortage of water, one of the basic problems of living on small islands that so obsessed Lescarbot he condemned all such sites as unsuitable for human habitation.

The lack of water for the quarantine station, which had troubled Moses Perley in the mid-nineteenth century, was still a problem at these hospitals in the early twentieth century. As late as November 1904, Mr. Thompson, supplier of water to Partridge Island, expressed his dissatisfaction (through the *Saint John Globe*) with the meagre pay he received for the troublesome task of delivering water from the mainland, in the *Neptune*, and pumping it up to the hospital. Even later, during and immediately after World War II, water piped to Partridge Island from Spruce Lake did not always reach the island. Pumps and pipes broke, which left the soldiers and civilians without this most basic necessity.

However, island dwellers during more peaceful times—our own included—have seemed surprised to learn that water has ever been a problem there. Among these are Noel and Jean Justason, residents on the Partridge Island light station as recently as 1980. "Oh, water was no problem," Mr. Justason stated, brushing the question aside. "A good well provided all we needed."

Such contrasting experiences and assertions about conditions on Partridge Island, voiced from one era to another, have provided food for thought. In few other places so small have judgments of residents and transients concerning the quality of life fluctuated so wildly. The questioner who is left mulling over the remaining fragments of evidence—the conflicting statements from the past, a shrunken skull on the lighthouse shelf, a sterling silver pepper shaker unearthed near the light station, an arsenal of broken weapons—can only note, inconclusively, that there are many versions of life on Partridge Island, each one largely a matter of opinion based on personal experience.

Notes on Partridge Island

1. It should be noted that Gesner's book was originally published in Saint John in 1842. Even so, it is surprising that he did not look more closely into immigrant quarantine procedures at that time. Since he did not do this, and since he presumably did not update his book before its 1847 publication in Britain, one can only consider this fact to be one more bit of evidence indicating that medicine never did interest Abraham Gesner much; his first concern was always geological formations and natural resources.
2. At the end of the eighteenth, and for the early part of the nineteenth century, seal oil was burned in the Partridge Island light for eight months of the year. During the remaining four months—the coldest ones—porpoise oil had to be used.

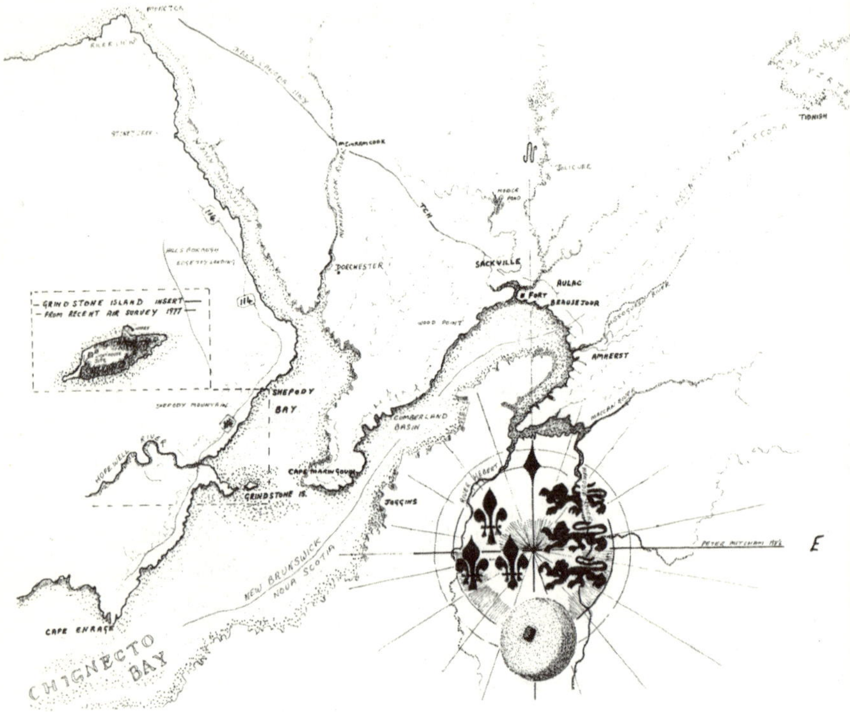

Grindstone Island in the upper Bay of Fundy

Grindstone Island

GRINDSTONE ISLAND. A small island lying off the south-eastern coast of New Brunswick, Canada, at the head of the Bay of Fundy. It has a number of sandstone quarries from which a fine quality of sandstone is exported, chiefly to the United States, for the manufacture of grindstones.

–The Encyclopedia Americana (1938)

Few small islands in the world have been considered significant enough to merit an entry in the *Encyclopedia Americana,* and this mention of Grindstone Island's chief export over many years calls attention to the exceptional quality of the stone that was quarried there, and to the value placed upon it in the outside world. This export, together with the island's significant location at the head of the Bay of Fundy, has been largely responsible for both the purgatorial and paradisal episodes in its history.

This island's size, now less than 2 kilometres long, about 550 metres across at its widest, and reduced to some 30 acres (from 113 acres when it was granted to the Anglican church in Sackville in 1823), has had no proportionate relationship to its economic importance, its scientific and religious interest, or its immense impact on the lives of the individuals closely associated with it. In fact, considering Grindstone Island's inconspicuousness on the map of the Bay of Fundy and the way in which this tiny block of sandstone has been eaten away over the centuries—both by quarrying and by erosion—its true wonder is that it still exists. This island's several names have reflected the uses to which its stone was put for more than two centuries. In Acadian times it was labelled "Île-aux-Meules," which means Millstone or Grindstone Island, an appellation that not only indicates that grindstones were discovered there in the seventeenth century, but that also underlines the importance attached to their being discovered there at that time.

Since one of the first priorities of the early Acadian settlements was a mill for grinding flour and cutting lumber, locating a reliable source of millstones or grindstones was essential. Because the new communities at the head of the Bay of Fundy ("la Baie Françoise" in those times) were a considerable distance from the potential supply source in the mother community of Port Royal, having infinite quantities of the right sort of stone near at hand would have dispensed with a good many worries. In any case, there had never been any assurances that Port Royal would have been able to provide the essentials for developing new communities when they were needed, as is evinced by the fact that the mills for both Beaubassin and Chepody had to be bought from New England traders. Although the French authorities tried to discourage such dealings with the English colonies to the south, they were unable to prevent transactions of this kind because they could not make any realistic promises concerning the prompt arrival of necessities from France.

Île-aux-Meules lay close to all three Acadian settlements at the head of the bay. It was only just out of sight of Beaubassin, about thirty kilometres across the water (now the Cumberland Basin); and was clearly visible from Chepody (now Shepody), about five kilometres across Chepody Bay. Les Mines (the Minas Basin settlements, dominated by Grand Pré) could also be quickly reached when wind and tides were right. However, the question still remains as to how the Acadians first learned of this useful stone, since even today Grindstone Island is not easily accessible and seems inhospitable. It looms impressively out of the fog—a castle island, at a distance not unlike Saint Michael's Mount off the Cornish Coast, or Mont Saint-Michel off the French coast.

The cliffs of Grindstone Island are nine to twenty-four metres high when exposed at low tide, except in one spot at the northern end where there is a small sheltered cove. The landing place that this cove provides allows an ordinary fishing boat no more than half an hour's leeway if it is to cast off before being stranded till the next tide. Given these conditions, one would suppose that few people of any age would choose to spend twelve hours (many more if fog rolled in or a storm came up) in a place which, at first glance, appears to have so little to offer. Still, whatever the reason, someone must have arrived on this island during the seventeenth century and spotted its potential usefulness. There are several explanations as to how this might have happened.

Early in the seventeenth century, when Champlain, de Monts, and Lescarbot were busy exploring and charting the coasts of Acadie, the King of France sent Sieur Prévert de Saint-Malo to look for precious minerals along the shores of this new land. Several of Sieur Prévert's investigations were focused on the Fundy coasts, particularly on the rock formations around the Minas Basin. The name "Les Mines" originated with these and other early seventeenth-century explorations for minerals. Sieur Prévert may have gone farther up the bay on occasion, and, after spotting the island then called Saint Louis, have put in there.

Stranded in this out-of-the-way place between tides, he could hardly have avoided viewing the island strata as well as the reef formation of abrasive rock just offshore. With the practiced eye of a geologist, he would likely have recognized the potential of the strata as a source of millstones. Although stone that served so mundane a purpose was hardly the object of Sieur Prévert's quest, it is nonetheless probable that, had he stumbled upon so useful a commodity, either he or members of his crew would have eventually mentioned their discovery.

Île-aux-Meules first appeared on the 1686 Franquelin de Meulles map. De Meulles, then intendant of Canada, has been credited with naming the island, although its spelling and consequent meaning, as W. F. Ganong emphasized, is not to be confused with de Meulles's name. De Meulles had visited Acadie the preceding year, and his first stopping place had been across the bay from the island at the seigneurial manor of M. de la Vallière. De Meulles's enthusiastic account of this visit still survives.

Apart from being gratified by the obviously thriving state of the new seigneury and its inhabitants, what had particularly impressed Intendant de Meulles was M. de la Vallière's foresight in meeting his ship on the Gulf of Saint Lawrence coast (at Baie Verte) and conveying him promptly on horseback across the isthmus to the comfortable and prosperous seigneury at the head of la Baie Françoise, thus sparing him the long and arduous journey around Cape Sable to Port Royal. Presumably, it was this journey that caused him to remark on the relative ease with which a waterway could be dug linking the Gulf of Saint Lawrence and the Bay of Fundy.

Once ensconced at Beaubassin, de Meulles not only enjoyed the comforts to which he was accustomed at home, but was assured that the boat he needed in order to proceed to Port Royal was ready to sail as soon as he was prepared to travel. Given the thoughtfulness and thoroughness of de la Vallière's preparations for his guest's comfort and convenience, it is little wonder that de Meulles was delighted with the first stage of his visit to Acadie. In fact he prolonged his stay over the winter. Not only did he not leave de la Vallière's seigneury in the autumn of 1685, but he did not set off from Beaubassin until July 1686 when he went to the mouth of the St. John River.

We do not know whether de Meulles actually went to Île-aux-Meules during this sojourn at the head of the bay or whether

Franquelin, the superb cartographer he had brought along, merely described it to him while he and his host sipped their wine over dinner or strolled atop the hill at Beauséjour. From this hilltop he could have looked out over his friend's rich marshlands and across the water to the impressive headland of Maringouin, which hid the island of which Franquelin and de la Vallière spoke. Still farther down the bay, de Meulles must have glimpsed Cape Enrage, its outline misty in the Fundy twilight. Perhaps de la Vallière recounted how many of the men whom he had brought to the seigneury during the past nine years had visited the island across the bay when fishing and trading with the Mi'kmaq. He may also have shown his guest samples of the island stone—an action that could easily have convinced the intendant that "Île-aux-Meules" would be a suitable name for this island.

Very likely, de la Vallière told his important guest of the innovative and laborious technique employed to cut grindstones from the island's reefs; a feat, which he doubtless pointed out, was only comparable—in skill, imagination, and strength—to the dyking of the marshes on the mainland. He must have described the channelling into the reef at low tide, the careful attaching of chains and buoys to the slabs that had been separated, the fastening of the chains to buoys that were later transferred to the wide-beamed fishing boats and the subsequent hauling of the slabs onto the prepared work beds near shore on the oncoming tide, and finally the wait (during which the men often went shad fishing near the island) for the slabs to be exposed again so that they could be shaped precisely with axes and chisels. He doubtless stressed the skill and speed with which the men worked to complete their task before the return of the tide, and the strength of their muscles as they heaved together to position the shaped millstones for loading onto the boats before the tide could overtake them.

After this 1685–86 visit of de Meulles, another thirteen years went by before the marshes on the side of the bay nearest the island were settled. In these years, the only European inhabitant of

Acadians cutting grindstones in the seventeenth century

Chepody Bay was, according to W. F. Ganong, a French priest who established a mission for the Mi'kmaq on Île-aux-Meules. Nothing else is known about his sojourn there. Then, just before the end of the century, a practical man, a miller, decided to settle the marshes opposite the island. He was Pierre Thibodeau from Port Royal, a successful man with a grown family who had acquired the nickname of "le meunier de la Prée-Ronde." Although well on in years—sixty-eight in fact—he was lured up the bay by all he had heard of the new territory. Upon seeing it, he was immediately convinced that it was all he had imagined. What could be better?

Here at Chepody were marshes that resembled those he was used to at Port Royal; and here, too, as a special bonus, off the island which was to be part of his domain, lay stone suitable for cutting into first-rate grindstones for his mills. He fixed on a knoll overlooking the marshes, the bay, and the island as the ideal site for the house he was to build from the oak and maple forests on the hillsides behind. Leaving several of his sons in charge of the new property, he returned to Port Royal to make arrangements to get together the materials needed to lay the foundations of the elaborate estate he envisioned. He reckoned that within a few years, and with the help of his sons and several other stalwart young men, he would have a seigneury comparable to that of M. de la Vallière across the bay.

As it turned out, this was not to be. As soon as he heard of Pierre Thibodeau's efforts to establish a new colony at the head of the bay, M. de Villieu, son-in-law of M. de la Vallière, decided that he ought to own this land himself. Since the land was located not far from his father-in-law's estate at Beaubassin, he argued that the newly proposed seigneury should be considered an extension of de la Vallière's. And who, he demanded, would be more fitted to control it than his son-in-law? M. de Villieu, in his notoriously strident and disagreeable way, insisted on his right to the land, whereas Pierre Thibodeau, though never really doubting the acceptance of his own claim, nevertheless calmly and good-humouredly continued to push for official recognition of his rights with the help of his own son-in-law, an educated and established resident of Port Royal. Pierre Thibodeau's faith in the acceptance of his petition had also been strengthened by the knowledge that his claim differed very little from that of his acquaintance, Mathieu Martin, who had been granted a comparable seigneury.

Pierre Thibodeau did not sit around waiting for a piece of paper to validate his undertaking; there was no time to waste. In short order he had a large house on the knoll built for himself, some of the marshes dyked, and barns and smaller houses constructed for his sons. Already the new colony was thriving. The powers-that-be seemed bound to reward such industry and vision.

Thibodeau and de Villieu waited nearly two years before a decision was made in France. When the news reached Port Royal in 1700 and was eventually relayed to Chepody, Pierre Thibodeau was stunned: all of the land he had staked out—marshes, woodlands, and the island—had been granted to M. de Villieu, as had the mills he and his sons had built on the streams that flowed into the Chepody River. Pierre Thibodeau could never have imagined how insignificant the claim of an upstart miller, whatever his achievements, would have looked to aristocratic decision-makers across the ocean compared to that of M. de Villieu, son-in-law of the respected M. de la Vallière, a gentleman long since celebrated in official reports.

Understandably, Pierre Thibodeau never really recovered from his disillusionment over losing title to the Chepody seigneury. He began to show his age, and died within several years at Port Royal. The pioneers at Chepody carried on despite these events, though much of the impetus for new development had died with its founder. The tiny settlement flourished and its inhabitants were joined, from time to time, by new settlers from Port Royal. The largest group arrived in 1713 to escape the British domination of most of the rest of mainland Nova Scotia, which had resulted from the signing of the Treaty of Utrecht. After this, the outside world appears to have forgotten Chepody for months at a time.

The reason Chepody was temporarily forgotten was because the focus was now chiefly on what was left of the old Acadie: Île Royale (Cape Breton), where the Fortress of Louisbourg was the centre of official concern for both English and French, although for different reasons. After the Treaty of Utrecht (1713), the French government spent an immense sum of money, which it could ill afford, maintaining the fortress and its large military and civilian population. The English were kept busy policing the old Acadian settlements of Port Royal, Minas, and Beaubassin, trying to prevent the flow of food from these rich farmlands to the fortress, where few crops could be raised on thin and rocky soil under an almost constant blanket of fog.

The battle of the Petitcodiac

Given this hostile atmosphere, the Acadian settlers at the head of the bay eventually saw the need to do something about providing for their own defence. Consequently, the residents of Beaubassin constructed a small fort—Beauséjour—on a height of land which was surrounded on three sides by water at high tide; and later the residents of Chepody built a much smaller fort, hardly more than a lookout post, on the point (now Mary's Point)—the land nearest Île-aux-Meules—in the hope that lookouts posted there could forewarn the settlement of potential enemies sailing up the bay.

Despite these precautions, the Chepody settlement had little warning before it was attacked in 1755, not long after the completion of the "fort." The inhabitants may not have even seen the English ships sailing up to Beauséjour some days before, or have had news of the battle and consequent English victory there. Whatever their information about the settlement on the other side of the bay, they apparently had no inkling of what was in store for them until the English ships rounded Cap Maringouin and dropped anchor in the five-fathom hole just off the north end of Île-aux-Meules, the only place deep enough and close enough to the settlement to afford suitable anchorage for large vessels. Within moments the English commander, Major Frye, had ordered his men into the ships' boats.

The farmer-fishermen of Chepody, seeing themselves hopelessly outnumbered by Frye's three hundred soldiers, gathered their families and fled, leaving their possessions behind. The soldiers destroyed the settlement at their leisure, setting fire to the buildings and butchering many of the animals. Then they were ordered back to their boats so that they could pursue the fleeing Acadians up the Petitcodiac and destroy any other habitations they chanced upon upriver.

When the invaders reached the village of Petitcodiac (now Hillsborough) they were ambushed. Boishébert and his resistance fighters, warned of Frye's intentions, had crossed the Petitcodiac River in their canoes at the bend where Moncton now stands, and had made their way downriver until they spotted some of Frye's boats anchored off the village. Stashing their canoes in the long marsh grass at the edge of the water, they crept through the chest-high grass and concealed themselves behind the dykes, which the English would have to cross on their way back to their boats. Boishébert and his men caught the soldiers off guard. In the ensuing struggle forty-five English soldiers were killed, sixty were seriously wounded, and many others were drowned as they attempted to get back to their ships. Only one of Boishébert's men was killed, but several were wounded.

With the English at the mouth of the Petitcodiac River briefly in disarray, Boishébert had bought sufficient time to get about thirty of the most vulnerable families to the fort on the St. John River, where, for a short while, they thought themselves safe. The strongest stayed in the woods along the Petitcodiac for nearly three years, struggling to survive and to wage a guerilla war at the same time.

The Newcomers

The Acadians never did get their Chepody lands back. American, British, and German immigrants took up where the Acadians had left off. However, the essential place names did not change greatly. Chepody (a name of Aboriginal origin) was anglicized to Shepody, and Île-aux-Meules translated to Grindstone Island. Like their predecessors, the newcomers cut the hay, pastured their stock on the marshes, maintained mills on some of the tributaries of the Shepody River, fished on both sides of the island, and quarried stone at Mary's Point and on the island.

Among the earliest and most prominent of these settlers were those who came from Baltimore, Maryland—people of German extraction who arrived about a decade after the Expulsion of the Acadians. Two companies were behind this land settlement scheme: one was dominated by General Haldimand, who was later (between 1778 and 1785) to become Governor General of Canada; the other by Thomas Calhoun, who in 1760 was appointed by the Crown as Indian agent in Nova Scotia, which then included what is now the province of New Brunswick.

Haldimand and his partners—Hoops, Hessencleaves, and Wallace—acquired a large grant in what is now Albert County, due in large measure to Haldimand's connection with the famous Colonel Joseph Frederick Wallet DesBarres, whose origins and career in a good many ways parallel Haldimand's. DesBarres was in a position to assess the Shepody lands because of his map-making activities in this vicinity. Indeed, he was so impressed with the land at the head of the bay that he set about acquiring twenty thousand acres for himself at nearby Memramcook.

By about 1768, much of the attention of both Haldimand's and Calhoun's companies was focused on the grindstones, which could be cut off Grindstone Island and the adjoining mainland and shipped to the United States for a handsome profit. Hugh Wallace, one of Haldimand's associates, remarked in a letter to Haldimand in 1768 that "the grindstone [is] an affair of consequence...." Both promoters seem to have wasted little time in establishing a lucrative business shipping grindstones from Grindstone Island to New York. The stones were rafted from the island and up the Shepody River where they were loaded onto large vessels.

This trade became so successful that soon others were eager to share in the profits. For instance, in about 1770, Thomas Calhoun's brother, William, sailed up to Shepody from Baltimore in a sloop that the brothers had decided to use for their trading ventures. All apparently went well until the spring of 1772, when they were loading their sloop for Baltimore with grindstones from Grindstone Island. They had transported small loads from the island to the sailing ship, which was anchored in the river, and on one of these trips from the island their boat was swamped; both William and Thomas were drowned, along with a brother-in-law named Peck.

As the years passed and ships of local manufacture as well as those from the outside world travelled up and down the bay, a busy trade in timber, fish, and stone was firmly established. And because of the ideal location, the bounty of nature, and the industry of the people, the new community prospered as the old one had. At length, in the way history has of repeating its patterns, a new geologist (an English-speaking one this time) was sent to assess the mineral potential of this part of the Fundy coast and its offshore island. His name was Abraham Gesner.

Born on the Nova Scotia side of the Bay of Fundy (at Cornwallis), Gesner had been an avid rock collector from boyhood. By 1836 he had already established a reputation as an authority on the minerals of Nova Scotia with his book *Remarks on the Geology and Mineralogy of Nova Scotia*. Not long after this, he moved to New Brunswick where he became this province's first geologist. In this capacity, Gesner set about assessing the mineral resources of New Brunswick, publishing accounts of his findings yearly between 1839 and 1844. On several occasions throughout these years he visited Grindstone Island.

Although he found nothing new on the island of immense commercial value—nothing comparable, for instance, to the rich find of albertite in mainland Albert County—he came up with an important scientific discovery. The island's cliffs, he noted, were an exceptional source of extraordinarily well-preserved plant fossils that clearly told a fascinating story of prehistoric life in that part of the bay. The present cold and blustery climate at the head of the bay, where spruces and firs struggled to cling to the soil in the teeth of wild gales, had once been tropical, and had favoured the luxuriant growth of palms and giant ferns. The evidence of the cliff face was incontestable. In one report Gesner observed: "...the remains of a variety of small plants, and leaves resembling those of the palm, may be collected from almost every rock on this [northern] side of the Island." Elsewhere he explained at greater length:

> The characters of the fossil animals, and more especially the flora of the carboniferous period, lead us to infer that the waters of the oceans and lakes, and also the climate were much warmer, and less variable than that at latitudes where coal strata are abundant. The existence of large true-ferns, which may be called the descendants of those now found in a fossil state, still growing in hot and humid climates, confirm the opinion that the temperatures of the atmosphere and oceans have been reduced in ocean latitudes.... In the steep embankment lie buried the lofty palm, the 'aurocarian' pine, that flourished in a moist and heated atmosphere; while on the surface, the hemlock and the hardy spruce of humbler growth, seem to mock the grandeur of their predecessors, over the graves of which they bid defiance to winter's snow and frigid vapours.

Although the fossil remains absorbed most of Gesner's attention on his various trips to the island, he did remark in passing that operations at the grindstone quarry had been discontinued "in consequence of running out." Those grindstones that were still being cut at several locations near the island were, he pointed out, "like those of the opposite shore in Nova Scotia...the best stones for cutting and polishing metals ever found in North America."

More importantly for the years that followed, Gesner observed that the strata would "afford excellent building stone of different shades of blue, red and chocolate color." His prognostications about the use of the island's freestone for building were realized in so short a time that even he must have been surprised. Some perspicacious individual closely connected with the Anglican church—the island's owner—must have read Gesner's report almost as soon as it rolled off the press and realized that the high-quality stone described might be just the material for crucial parts of the new cathedral that was then being planned for the capital city. The church's ownership of the island made the whole proposition seem feasible. And so, within the next few years, stone was cut on Grindstone Island and loaded onto boats, which carried the huge pieces down the Bay of Fundy and up the St. John River to the cathedral site. There, the columns, groins, and arches of island stone were (in due course) installed in Christ Church Cathedral, where they have contributed greatly to the special beauty of this Fredericton building.

Some of the men who cut the stone were immigrant Irish and Scots, trained at quarries in their native lands. With them worked local laborers, both anglophone and francophone, from the nearby mainland—men who had not yet established themselves in the farming, fishing, or shipbuilding enterprises that then flourished on the shores of Shepody Bay, the Petitcodiac River, and the Cumberland Basin. Because of the difficulties of getting to and from the island, a large boarding house was built near the landing place and within a few steps of the quarries. It was to this snug, wooden structure, lined with birchbark to keep out the gales and

Fundy fogs, that the men retreated, easing their aching muscles and bones with heavy doses of rum and a hearty hot meal; and it was here that they fell asleep, sometimes so exhausted that they did not even remove their boots or overalls. And despite the way they slaved, few managed to save enough money from their meagre wages to buy the land or the boat needed to begin another sort of life. Many were free men in name only. Records show that while the Anglican church kept close tabs on what transpired on the island, it did not usually supervise quarrying operations there, but leased the quarries to a series of entrepreneurs—some local, some foreign.

By 1855 the quarries on Mary's Point and Grindstone Island had become so important that they were the subject of the report delivered to the Boston Society of Natural History, and subsequently published in the *Saint John Evening News* on November 21. Parts of this report provide valuable information about the extent of the quarrying carried on at these places and the regard in which the stone was held in mid-nineteenth century.

> On reaching the mouth of Chepody Bay we come to Mary's Point, and Grindstone Island, where beautiful buff-colored, and light red sandstone of the Carboniferous group, are quarried for building stones, and a blue variety of the same formation which is used for making grindstones is found. The Mary's Point quarries present a scene of activity, owing to the enterprise of Mr. Archibald, who has employed upwards of seventy-five quarrymen, under the directions of Mr. George Lang, in quarrying out the freestone for architectural uses. These sandstones are found to be of excellent quality for building purposes, and are now being largely imported into the United States under the Aegis of the reciprocity treaty between the United States and Great Britain. Mr. Alger has demonstrated the durable nature of this sandstone, before this society, and our observations authorize us in confirming his opinion, that it withstands the action of the frost even in the climate of St. John.

> Grindstone Island is one and a half kilometres long, and one kilometre broad, and is estimated to contain about 140 acres of quarry stone. The average height of the island above highest tides is eighteen metres. In obtaining slabs of the blue sandstones for making grindstones, the tides perform valuable services, for the quarryman has only to split out the rock at low water, and fasten it by chains to his flat boat and allow the rising tide to float it, so that he can transport the stone to high water mark, and leave it where he can work upon it at his leisure. The best grindstones are obtained from the strata which are submerged at every tide....
>
> The extensive quarries are supplied with heavy and massive cranes, and a railway, for the removal of the stones to the wharf, where they are put on shipboard.

It was just past mid-century, at the height of the quarrying operations, that it became fashionable for the young people of Albert County to go across to the island for picnics. Here, on the threshold of their own quiet world, they could watch the quarrymen at work and observe the large ships from New York and Philadelphia take on cargo. Or, if on a Sunday or holiday all hands were idle, the young people could chat with men who had but lately walked the streets of the most important cities of the continent.

Before or after such exciting encounters, the picnickers generally walked around the island on the cliff-top path, the young men going in one direction and the young women in the other. When their paths converged, considerable merriment ensued. Yet, some of the repercussions from these pleasant and innocent outings were tragic and unforeseen. Now and then, a young person who had conversed with the strangers fell ill with cholera or one of the other often-fatal diseases that plagued the world's busy ports throughout the nineteenth century. Then a baby brother died, a mother, a father...and it would be years before anyone would connect these island visits and the illnesses that became endemic on the neighbouring mainland. Meanwhile, although the developers changed frequently, the island's extraordinary resource continued to be exploited at a steady pace. As long as the market was there, someone could be counted on to see that the stone was got out. After the boom years of the 1850s, a series of entrepreneurs presided over the stonecutters.

In December 1860 a five-year contract was drawn up between the churchwardens in Sackville and Alexander Rogers of Hopewell who, for a yearly fee of forty dollars, was to have quarrying privileges on the island. Later, in 1878, a Boston company, headed by Austin I. Roberts and John F. Roberts, leased the island. By this time leasing fees had more than doubled. For just a portion of 1878 the Roberts paid fifty dollars, and for all of 1879 one hundred dollars. Previous to 1878 the Roberts's company had been cutting stone on Mary's Point, the nearest point of land to Grindstone Island, but in 1878, when the Mary's Point quarry was no longer profitable, they moved the headquarters of their operations to Wood Point, supervising the Grindstone Island cutting from there.

Written into contracts for all these leases was the stipulation that "the privilege of getting grindstones under the tide and also the fishing privilege belong to Saint Ann's Church, Sackville." Indeed, with the earlier of the above-mentioned agreements, Mr. Rogers's activities on and around the island were most severely curtailed. Not only was he not allowed to set nets, or establish weirs on the flats around the island, or to cut grindstones on the offshore reefs, but the lease even stipulated that he was not to camp on the island. Such rigid restrictions posed a real hazard to fishermen of the whole region. For them—particularly for those who had almost a two-hour sail down from Wood Point—the island provided a safe place to wait out a storm. Without being able to use this sanctuary, their lives were more frequently threatened than they would otherwise have been.

Throughout the decade following Abraham Gesner's explorations of the rock formations of Grindstone Island and the adjoining mainland, and while quarrying was in full swing (from the mid-1840s to the mid-1850s), the priest missionary at Memramcook, Father Ferdinand Gauvreau, crusaded for changes in the methods of exploiting the important shad fishery that radiated from this strategic island. "The shad-fishing ground," he reported to Moses Perley, "extends from about eight miles below the Bend of Petitcodiac, at Stoney Creek, as far down as Grindstone Island, in the Bay, and thence easterly into the Bay of Cumberland...." Father Gauvreau was particularly concerned with the depletion of this fishery due to the increasing use of weirs and stake nets on the Shepody flats opposite Grindstone Island as well as in the Cumberland Basin. Though recognizing that such methods had been traditional in these locations from the time of the seventeenth-century Acadian settlements, Father Gauvreau could nevertheless only see them as the cause of the waste of an important resource on which his parishioners—some of them descendants of the original Chepody settlers—depended.

Many of the fish caught in the weirs and stake nets on the incoming tide were killed and then swept out into the bay by the receding tide, useless, save as food for the marauding sharks (the two- to three-metre *carchaias vulpes*—or fox shark—and dogfish). Father Gauvreau, himself an expert fisherman and an exceedingly practical man in general, set out by example in his own boat, the *Saint Peter,* to persuade both his own parishioners and others to use drift nets on the river and bay. By using these nets—set out from small boats that drifted in and out with the tide, day and night, mostly between seed time and haymaking—enough fish swam free to provide for future generations. Moreover, because all of the fish caught in the drift nets could be hauled in as the nets filled, there was little or no waste.

For such an operation, a good boat and sails was reckoned to cost from £10 to £12 at this time, and Father Gauvreau, in a letter written to Moses Perley in December 1850, announced the success of his years of persuasion, stating that since using this technique "the average quantity of barrels of shad caught in the upper part of the Bay, among the French Acadians, is from 1,000 to 1,500 a year... this year, 100 French fishing boats have been drifting down the Bay, all fine and strong boats, sailing well, fully prepared for any storm."

There were, by Perley's reckoning, about the same number of small English craft congregated in this relatively small stretch of water. All these little boats, together with ships of large tonnage coming up and down the bay to transport stone from the quarries and produce of the fertile earth and sea to markets down the eastern seaboard, made the head of the bay a busy waterway. In his *Handbook of New Brunswick,* Moses Perley provided a good deal of specific information to indicate just how much traffic there was:

> In 1853, eighteen vessels of 1,328 tons, entered at Sackville; ten vessels, of 771 tons, entered at Dorchester; and twenty-one vessels, of 1,646 tons, entered at Moncton. Vessels of all sizes, up to 1,000 tons, are built at each of these places.... In 1835, thirteen vessels, of 1,401 tons, entered at Hillsborough. These vessels carried the various products of this country to places abroad; and there were many coasters also employed in carrying produce to the Port of St. John.

With all this prosperity, all these comings and goings, the great scourge of the era: cholera, appeared in Shepody. In personal diaries of the time, such as that of Eliza Ann Rogers of Hopewell Hill, which covered a period from April 17, 1852, to February 1896, there are (particularly in the 1850s) repeated references to the many vessels "coming up the Bay," and to the cholera deaths in the community. So frequently did Eliza's neighbours—particularly young children—succumb to this plague that her reports of death from this cause are generally brief and matter-of-fact. Only occasionally did she comment, and then only when the circumstances were unusual; as on January 19, 1853, when she observed: "Mr. Tucker had his child dug up and its likeness taken, then buried again."

Marvin Snowdon's fishing boat in the cove on Grindstone Island in 1985. The landing place which this cove provides allows an ordinary fishing boat no more than half an hour's leeway if it is to get off before being stranded till the next tide.

Cholera was not the only pressing danger to be reckoned with at the head of the bay. The pack ice in the spring and the heavy fogs throughout the shipping season were notorious marine hazards. Given the increasing numbers of foreign ships heading up the bay and the steady numbers of fishermen "night drifting" down, a light in the most strategic position seemed increasingly important. Grindstone Island was the logical place to build such a light station.

In response to this generally articulated need, the government, in June 1858, began negotiations with the church in Sackville for a half-acre of land on the southern side of Grindstone Island for the "purpose of a lighthouse." By March 14, 1859, the churchwardens had agreed to lease or sell this piece of land for £50. During the 1859 shipping season, the lighthouse was in operation.

The light station chiefly benefited shipping from away, since local captains were well aware of the idiosyncrasies and the positions of reefs at the head of the bay. And indeed, shipping records indicate that the vessels that did eventually founder near Grindstone Island or on the journey there, tended to be either ships of foreign registry or those whose home ports were not close by. For instance, in 1899 the barque *Ingomar,* of Norwegian registry and out of West Hartlepool, Great Britain, was wrecked in the fog just southwest of Grindstone Island; in 1895 the barque *Equator,* sailing from Rio de Janeiro to Grindstone Island, ran aground in fog off Long Island (off Digby Neck); and in 1913 the *Gerald Turnbull,* of Welsh registry and sailing out of Cardiff to Grindstone Island, also ran aground near the entrance to the Bay of Fundy—at Gannet Dry Ledge. These were all large and valuable vessels of 1,182; 1,228; and 1,967 tons respectively. Obviously the trade in stone necessitated the use of ships of considerable tonnage.

Yet perhaps what most impresses the modern investigator of such facts is the realization that entrepreneurs from so far away as Rio and Cardiff would even consider sending ships all the way up to the head of the Bay of Fundy for a cargo of stone. It seems hard to believe that businessmen in these places would have chartered vessels to sail to an island that they must not even have been able to locate on a good many of their maps. Still, not all the vessels that got into difficulties on journeys to or from the head of the bay were captained by foreigners. Even ships of Fundy registry skippered by Fundy men sometimes came a cropper on such voyages. One such vessel, which was wrecked on Grindstone Island, was the brig *Annie Bogard.* Her port of registry was Digby and her master was from the famous family of Digby county mariners, the Slocums.

On December 22, 1881, the small (148-ton) *Annie Bogard* had taken on a cargo of coal at Dorchester Island and set sail for her destination, Newburyport. The weather was uncertain, but Captain Slocum knew that if he was to get his cargo out before the river and the head of the bay froze for the winter he could not afford to dally. However, he was no sooner out of the river and into Shepody Bay (on a clear day he would still have been within sight of Dorchester Island) when the storm that had been threatening, broke. Slocum attempted to reach safe anchorage at Five Fathom Hole so as to ride out the storm, but before he could, the vessel was driven onto the reefs extending from Grindstone Island.

Slocum and his crew of five, hopelessly stranded with the vessel fast breaking up, lit torches as signals of distress. The lightkeeper, John Stiles, spotted these signals almost at once and hurried down to his boat at the cove. With Allen Fales of Hopewell and two other men, he put off for the distressed ship. But no sooner had they got out of the cove than these two, considering the rescue mission doomed, decided to opt out and forced Stiles to turn back to land them. This loss of time appears to have had a good deal to do with the tragic outcome of the venture, since several minutes turned out to mean the difference between life and death. When Stiles and Fales reached the brig it was already on its beam-ends, nearly submerged, the masts almost horizontal, and all six men clinging to one mast. Captain Slocum, the mate (McLean), and one sailor dropped into the water and were picked up, but before the others could follow suit, the mast went under, carrying all three down with it. Stiles, like other lightkeepers on remote Maritime islands, seems to have considered this heroic rescue all in a day's work. From the beginning, the ties of Grindstone Island's lightkeepers were, understandably, with the settlements along the nearby Shepody coast in Albert County rather than with Westmorland and the church that owned the island.

The wharf at Harvey was the most convenient place from which to launch a boat to the island, and several places along the Albert County shore were natural lookouts for mainlanders concerned with the functioning of the light station and the safety of its inhabitants. The mainland residents of several strategically situated homes watched for trouble on the island as carefully as the keeper attended to his light and to mariners in distress. With powerful telescopes—mostly family heirlooms that had once been employed on the decks of sailing ships—some Shepody residents could see the keeper and his family moving about the field in which the lighthouse stood, and in turn, the keeper could identify people out in their yards at Hopewell.

The Last Lightkeeper

Accidents and severe illness seem to have been rare on the island, at least during the twentieth century. Some of the keepers and their families completed a twenty-year term without ever once signalling for help. This was the case during the tenure of the last lightkeeper, Mr. Wainwright Weston, who, with his wife and five children, ran the station from 1950 to 1970. However, now and then in the years before the Westons' occupancy, accidents did happen and help had to be summoned.

For instance, on one summer's day in the early 1930s, the keeper's assistant was soaping the kerosene-run motor belt. Before he knew what had happened, one of his sleeves caught in the machinery and he was whipped over the motor, smashing his chin and jaw. The keeper hastened to the post and quickly raised the distress flag. Not many minutes later, his young nephew, peering through the telescope, spotted the shad boat putting out from Harvey and heading towards the island. In little more than an hour the injured man was ashore and on his way to the doctor.

Nature herself seems to have intended this relationship between the island and the Shepody shore—the shoal stretching out from Mary's Point to Grindstone Island indicates this. At very low tide the water is sometimes only up to a man's shoulders. Knowing this, several local men, just after the turn of the last century, decided to walk the kilometre to the island from the point, pushing a small boat ahead of them in case they lost their footing. For the lightkeeper who they visited upon reaching the island, this crossing was a memorable undertaking, and the source of much discussion at the time as well as for years afterward.

A shallower, better crossing place is in a direct line from Hopewell Hill to the island. On an exceptionally low tide the water is only about a half-metre deep. Even so, it is not a walk to be undertaken casually. A five-kilometre wade through several metres of icy water with no firm footing in the soft mud and a time limit imposed by the incoming tide (the world's highest) is not most people's idea of a pleasant excursion. Mr. Weston remarked that he had never made the crossing in this way, though he had contemplated it several times. He knew that cattle had, at one time, been driven across here without great difficulty and that riders on horseback had found the crossing easy when the tide was right, but that pedestrians had been less fortunate.

One man, walking out across the flats towards the island, had barely survived. He had gone to look at a neighbour's weir when he noticed a slight rise in the level of the water. He turned around at once, trying to hurry, wanting to run. But he couldn't run in several metres of water, particularly with several centimetres of oozy bottom sucking at his feet with every step. In this nightmarish race with the tide, the man's shoulders soon felt the water's paralyzing embrace. He lost his footing altogether, but miraculously, with a last wild thrust of his body in the surging tide, he was swept in to shore and was able to scramble up into the long marsh grass to safety. The account he later gave of his ordeal did not tempt others to repeat the experience.

Lightkeepers generally, and those on Grindstone Island in particular, have demonstrated little inclination for such dicey confrontations with the elements. This is not really surprising since they have generally been hired precisely because they possess attributes that are not likely to precipitate them into foolhardy undertakings. An ability to plan ahead, an even disposition, a practical bent, and an aptitude for coping alone with both routine and unforeseen circumstances are all prerequisites for a satisfactory lightkeeper. Since these enduring qualities also tend to make for a satisfying life anywhere, one begins to understand why it is that, by and large, island lightkeepers and their families seem to lead exceptionally balanced and fulfilling lives.

The very fact that the island circumscribes their lives becomes, paradoxically, a positive element, since its very limits make it manageable and understandable in its totality. Thus the island lightkeeper and his family, early in their island stay, must come to terms with their defined environment and with their own existence within it. If they can manage to do this, their new kingdom becomes an extension of themselves and they of it. For a good many lightkeepers and their wives, their time of sadness and unbalance comes only when they have served their time—when they must leave their island. Returning to the mainland world, they are sometimes as uneasy as Adam and Eve must have been after being expelled from the Garden of Eden.

To compare life on Grindstone Island with that in the Garden of Eden may seem, on first consideration, far-fetched. A lonely light station set up in a bare field, exposed to tempestuous winds funnelled up the bay, and flanked by closely drawn-up battalions of dark spruce, seems hardly the proper setting for a pastoral idyll. Yet, such a tranquil rural interlude has been lived there. The high calibre of its inhabitants has not, alone, been responsible for the superior quality of life on this island. The many positive aspects of the natural environment should not be underestimated.

Judged by every criterion for maintaining a high standard of island living, Grindstone is exceptional. It provides the essentials in abundance. The water, which bubbles up from a spring on the south end of the island, has been pronounced, by residents and visitors alike, the best they ever tasted. Wood for fuel and building is available in large quantities, since most of the island is wooded, with hardwoods dominating the protected interior. The soil is excellent for gardening and pasture. Game animals, both deer and moose, which periodically swim over from the mainland, have inhabited the island from time to time, and fishing just offshore has always been excellent. Thus, although Grindstone Island may at first appear to be a somewhat formidable and inauspicious place when compared with a seemingly more hospitable and gentler island such as Dochet (Champlain's) Island, there has proved to be no comparison between the two islands' capacities for sustaining life— Grindstone comes off best on all counts.

When the Westons first arrived on this island in 1950 with their five children, the youngest was only two. As anyone who has coped with a family of active youngsters will understand, the Westons had no sooner unpacked and set their new establishments to rights than Mrs. Weston found that it was time to do a washing. Because there had been so many things to think of, she had not put the washing on the line until the afternoon, so the clothes were left out overnight. The next morning when she went to bring them in, the line was absolutely empty. As Mrs. Weston recollected: "The wind blew so hard all night that not a stitch remained. That was a real surprise! What a beginning to our stay! You can imagine that after that I learned to pick and choose my washdays and never to leave the clothes out all night."

"Well, at least the wind didn't serve us the way it did one of our predecessors," Mr. Weston took up.

> Before our time they had an outhouse perched up on the edge of the cliff. One night the keeper visited it when a considerable gale was blowing. He noticed that the little house was rocking somewhat in the wind, but didn't think too much about it since the wind on that exposed part of the island was always pretty brisk. So in he went and was just making himself comfortable when an extra powerful gust swooped down and shook the outhouse

> vigorously. Well sir, he didn't waste any time getting through that door and starting down the path. But he'd hardly taken a couple of steps when he heard a great roaring and bumping behind him. So he stopped and turned around just in time to see the outhouse sail up and whirl over the cliffside into the sea.

When the Westons moved to Grindstone Island they had brought with them, in the manner of much earlier settlers in remote places, chickens and a cow. The cow would become a family pet, growing so tame that she seemed to look forward to giving the children rides on her broad back. Yet it was this same creature, amiable and tractable though she was, that caused the only accident that occurred on the island during the Westons' entire twenty years there.

Near the lighthouse was an old cistern that had been used to collect rainwater for the resident family's general washing needs, thus providing soft water in addition to the drinking water, which they hauled up the cliff path from the spring using pail and a yoke. This cistern had been boarded over and forgotten early in the Westons' residence on Grindstone, when the old water system had been replaced by a well and pump. Since Bossy roamed at will around the meadow, and since she preferred to stay as close to the light station as she could, she happened one day to step onto the boards of the old cistern. Her weight was too much for them and gradually, as they gave way, she was deposited in six or seven centimetres of water some four metres below ground. Because her descent had been gradual she had not been hurt, but she was shaken. Nothing so dreadful had ever happened to her before in her entire life. Even her journey to the island in the first place—a sickening, pitching three-quarters of an hour in a fishing boat—had not been nearly as traumatic as her sudden descent into this pit. Bellowing at the top of her lungs, she protested her imprisonment.

Soon the family was ranged along the cistern-side, staring down at the wild-eyed and pathetic captive. But Mr. Weston—"Pappy," as his family affectionately called him—was not a man to stand around pontificating or wringing his hands. Quietly sizing up the situation, he called for canvas from the shed and thick ropes from the lighthouse. Then he and several of the boys climbed down beside the distraught animal, patting her to quiet her down while they made a sling of the canvas and secured it under her belly. Then up went Mr. Weston again to man the pulleys he had rigged above. In short order, Bossy was hoisted out of her prison and was soon back in the field munching contentedly, having promptly forgotten, it would seem, her recent ordeal. The family, however, was less able to forget an episode that became immediately part of a treasured family history.

At this time the Westons had another memorable pet, a raccoon that had been on the island when the family arrived. The raccoon first introduced himself by visiting their garden plots, which they kept tucked away in the woods out of the sweep of the wind. Sampling all the delicacies, seemingly provided for his enjoyment, he, in typical raccoon fashion, fixed particularly on the corn, appropriating most of it for his own use. But in the fall, once the garden was finished, the raccoon's curiosity overcame his timidity and he made regular visits to the field in which the lighthouse stood, apparently anxious to savor other delights that his human neighbours might provide. He was not disappointed. The children introduced him to bread and peanut butter. He was instantly won over. Although he had eaten corn enthusiastically, he found peanut butter infinitely superior. He became hooked on this new food, which he realized was only to be found in the vicinity of the house. Every day he came closer, forgetting caution in his single-minded quest for the food he craved.

One day he noticed the back door open and no one near; so, doubtless hoping to discover an immense cache of bread and peanut butter, he ambled in to have a look. The room he entered—the kitchen—did not have the right smell that day. Besides, he almost lost his footing on the hard, slithery floor on which his claws made a disconcerting scratching sound—worse than on ice. Then all at once he caught a whiff of the smell he'd been seeking. Scurrying across the rest of the slippery floor and through a large opening, he was soon in more comfortable surroundings. Here in the living room, his claws sank into a material almost as spongy as moss or grass. He followed the wonderful peanut butter smell up onto a couch where all at once he was face to face with two of the children. Each was holding a large slice of bread and peanut butter. They shared their snack with him.

After this the raccoon visited the Westons' house often. Year after year he appeared on the doorstep, until he became so old and blind that he could scarcely find his way. Increasingly, the Westons felt responsible for his well-being. And so, when the time came for them to leave the island, Mr. Weston arranged for the raccoon to be taken to the Magnetic Hill Game Farm in Moncton, where he would receive both the care he needed and the treats he craved.

Time never hung heavily on the hands of the island family. In the evenings they made music. As Mrs. Weston remarked, "Pappy and all the children were really musical, so we had our own music, lots of it, all the time." The music was such a success at home that it was not long before Mr. Weston had his own show on Radio Amherst and began hearing from fans all over the region. Eventually, some of the friends he made this way visited him on the island.

Although the Westons did not have a telephone they were in touch twice a day with the department of transport by two-way radio, and a helicopter stood by in case of emergency. The helicopter also transported inspectors to the island on their routine, though usually unannounced, visits. Because of the ice in the upper part of the bay during the winter and the consequent halt of marine traffic between late December and April, the Grindstone Island light station was closed down for this period and the Westons returned to their mainland home; a residence which, appropriately, looks out towards the south end of Grindstone Island and its dominant feature: the now defunct light station.

Sometimes the Westons were on the light station at Christmas, sometimes not. One year, when they had expected to spend Christmas on the island, the government boat arrived unannounced from Saint John to take them off. Quickly they got together everything they thought they would need before spring. They were back on the mainland when Mr. and Mrs. Weston remembered that they had left all the children's Christmas presents hidden under the bed in their room. In short order Mr. Weston returned to the island—this time in a small fishing boat—to collect the forgotten gifts.

During the four months they were off the island the Westons sent their children to school, finding that their education had not suffered in the time they had spent on the island doing correspondence lessons. This, too, was generally the time when the Westons went visiting. If they left their Albert County community it was, more often than not, to call on other Fundy lightkeepers, invariably those with whom they had been in contact on the light station's shortwave radio. One of their most cherished memories of this time is of a visit to the light station at Swallowtail, Grand Manan, where, the Westons admitted, "We felt right at home."

Yet apart from witnessing the treacherous Fundy fogs and storms, the keepers at Swallowtail and Grindstone light stations looked forward increasingly to dissimilar prospects. Although the design and size of the boats that plied the waters near Grand Manan had changed markedly during the more than a century since Moses Perley and others had observed the many vessels moving through the waters around this island, the numbers had not appreciably altered. In contrast, at Grindstone Island the small amount of traffic by mid-twentieth century bore no resemblance to the busy comings and goings in the mid-nineteenth century. Now hardly anyone ever remembered how, just a century before, grindstones quarried on the island had been common currency in Albert and Westmorland Counties; how, in almost any store, they could be exchanged for other commodities as readily as dollars. So, while the rest of the world operated on the gold standard, this part of the world could be said to have operated for a while on the grindstone standard.

Lightkeepers' names inscribed in the rocks on Grindstone Island

Even during their lifetimes, the Westons had noted a striking decline in the shipping at the head of the bay. In their childhood and throughout their early youth, they had watched the many ships, among them stately schooners, sailing up the bay, but in the twenty years at the light station on Grindstone Island they had seen few vessels apart from passing scows, a handful of local fishing boats, and, once, a beautiful black-hulled yacht. By 1970 the head of the bay on the Shepody side was almost as little frequented as it had been three hundred years earlier. A lighthouse here had become an extravagance.

Since the closing of the light station, the island has had no permanent human residents. Birds have now taken it over—gulls around the perimeter, herons (the fishermen's "cranes") in the interior. The buildings on the light station have rotted away and the path around the island, which the Westons kept cut and which had once been the favourite walk of courting couples, has now been almost obliterated by the new growth of trees and shrubs. Some of this path has also crumbled into the sea. Even the names inscribed on the island's rocks are growing fainter with every tide that passes over them. Yet here the contemporary visitor may still decipher portions of the old record—the names of other visitors and residents long since dead. And in a place apart, always covered at high tide, is the chiseled list of this island's keepers, each one the island's guardian for twenty years or more.

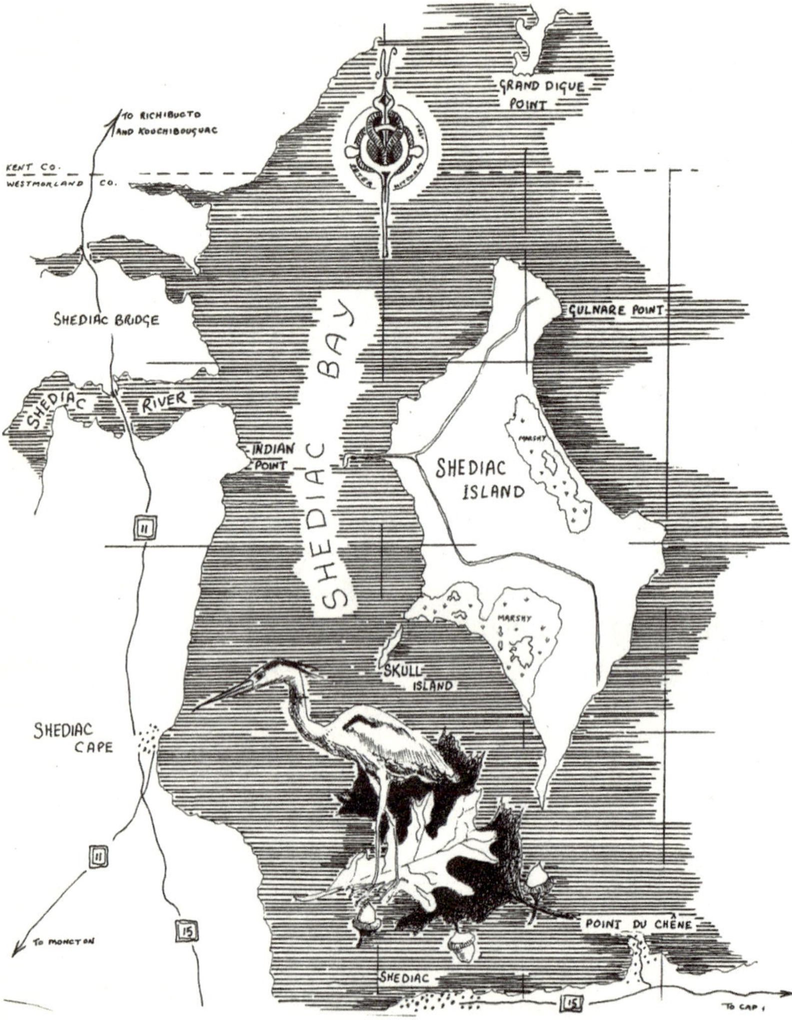
GRAND DIGUE POINT
TO RICHIBUCTO AND KOUCHIBOUGUAC
KENT CO.
WESTMORLAND CO.
SHEDIAC BRIDGE
GULNARE POINT
SHEDIAC BAY
SHEDIAC RIVER
INDIAN POINT
SHEDIAC ISLAND
MARSHY
11
MARSHY
SKULL ISLAND
SHEDIAC CAPE
11
15
TO MONCTON
POINT DU CHÊNE
SHEDIAC
15

Shediac Island

Shediac Island is essentially an overgrown sandbar. Stretching nearly three kilometres across the entrance to Shediac Bay, it is a beautiful and vulnerable spot. With its strong new growth of trees and abundant bird population, this island is a naturalist's dream come true. Without human inhabitants for many years, Shediac Island has reverted quickly to its natural state. Today, no trace remains of the dwellings of the seventeen families (sixteen of them Acadian), who were seemingly so permanently established here at the beginning of this century.

Apart from the fact that many of the trees are new growth, this island must not look so different from the way it appeared when Nicholas Denys first sailed by it in the seventeenth century, or when Governor Wentworth dropped anchor off it and DesBarres first put it on his charts in the eighteenth. Miraculously, even the oaks—that gave the opposite point (Pointe-du-Chêne) its name, and which have long since disappeared from there as elsewhere along the coast—have not died out on the island. Visiting this tiny, unspoiled kingdom makes one realize how breathtakingly beautiful and rich in life this whole stretch of coastline must have been before our forebears cut the pines and oaks to Governor Wentworth's specifications, destroyed hundreds and thousands of seabirds for food and fun, and ringed the shore with chains of wooden cottages. No wonder that Denys, some three hundred years ago, called part of the mainland and one of the islands not far above Shediac, "Cocagne," which translates into Utopia.

A close look at Shediac Island also alerts the visitor to the fragility of this particular environment and the vulnerability of its non-human population. "This island could disappear next week if people aren't more careful," a Shediac man who loves the island remarked thoughtfully. "It wouldn't take much. Just one of those campfires out of hand sweeping across the island, followed by a week or so of storms and high tides, could do it. The island's so low and there's no rock base. Only the trees anchor it."

Anyone who doubts the truth of these comments need not look farther than Little Shediac Island (sometimes called Indian or Skull Island) to realize that he is right. At low tide, this minute island is linked to the larger island by an umbilical cord, a thin sandbar. Despite its protected situation in the lee of the larger island, Little Shediac grows smaller every year. Long denuded of trees, its banks are steadily eroding. In a few years it will be washed away. Already the sea has swallowed what remained—at the turn of the last century—of an old Aboriginal fort here.

But why, you may wonder, did human beings abandon Shediac Island—particularly in view of its great beauty and proximity to the town? Partly for the same reasons they have left other offshore islands: they wanted the newly obtainable conveniences found only on the mainland. It was all very well to live on an island when mainland and island were, by our present standards, both inconvenient. Indeed, up to the early years of this century, islanders often thought themselves better off than their mainland friends and relatives. In summer the islanders' boats were obviously a much quicker, smoother, and cleaner way to travel than the mainlanders' dusty and bumpy horse-drawn carts or wagons; and in winter, everyone—on island and mainland alike—was snowed in or icebound. But once cars became commonplace, roads relatively smooth, and snowplows capable of clearing the winter roads, and movie theatres and hospitals were within reach of most mainlanders, islanders felt left out of mainstream life. They embarked for the mainland where electricity, the telephone, the railway, schools, hospitals, churches, and stores were all within reach.

But there were other reasons too. Shediac Island, like other Northumberland Strait islands, was an especially inconvenient—even dangerous—place to live when the ice was unsafe. Even one kilometre is a long way when shifting ice cakes or soft, grey patches of ice lie between you and your mainland destination.... One by one the families left the island. The school closed. A little girl, Annie Robinson, the lightkeeper's daughter, fell through the ice and drowned on her way to school on the mainland. A woman lay dying of cancer with three small boys to look after while her husband went across the ice for the doctor. The neighbours had already moved away. No one was nearby to lend a hand.

"It's all right to take a chance ourselves, but not risk our kids' lives.... It's time to move," was murmured across the island. The lighthouse, seemingly a fixture on this island, was shut down and burned—and with the demise of the lighthouse, that sustaining beacon of island life, the last islanders to consider Shediac Island home—the Clements and the Fougeres—packed their belongings, dragged the moveable parts of their dwellings across the ice to Pointe-du-Chêne, and struggled to put down new roots there.

Still, apart from all these difficulties, perhaps the greatest deterrent to human residency on Shediac Island has always been the mosquitoes and blackflies. The swamp and ponds that cover much of the island seem to provide these creatures with an ideal habitat. They are really menacing, more voracious even than their notorious arctic cousins, and the visitor is hard-pressed to imagine how the islanders managed to live with them for five months each year and maintain their sanity. A few hours at a time would drive most people to the brink of madness. Since these rapacious predators seem impervious to even the most-touted contemporary insect repellents, how the islanders coped with these pests so long before the invention of insectifuges is impossible to imagine. Former islanders tell of the priest's coming to exorcise these demonic creatures, and of ensuing periods of temporary relief. But the flies and mosquitoes always returned.

For the island's present inhabitants—birds of countless species—these clouds of insects are not pests. Indeed, they provide the birds with a wonderful supply of fast food. And when sated with these delicacies, herons, terns, and gulls join sandpipers on the shore or in the shallows for a fish fry. Day after day on the hot midsummer sand, the tiny caplin toss and glisten like smelts flipped in a frying pan. Every bird has more than he can eat. As one observant individual remarked: "It's the only place where I've seen gulls unable to choke down another morsel. I've seen them turn away from the caplin looking ill, like overfed travellers waddling away from an all-you-can-eat Howard Johnson fish fry."

Not all the island's birds are so gregarious and noisy as the feasting gulls, sandpipers, and terns, and not all are drawn daily to the crowded bar. Like their human counterparts elsewhere, the birds here differ greatly from one another in both habits and habitats.

Near shore, though seemingly as far away as they can get from the raucous crowds, family groups of scoters (coots) feed quietly and companionably. If an intruder disturbs them, with one accord they lift their sleek black bodies half out of the water and plane across its surface with the speed and finesse of accomplished water skiers. Once they feel that they have left the troublemaker behind, swamped in their wakes, they settle down again without so much as a backward glance.

More solitary and private is the kingfisher, glimpsed by chance as he dives for fish from the branches overhanging a clear pool, or the hawk who sits in the tall tree above the pond, seemingly meditating upon lofty matters. Yet it is the herons that dominate this island. One or two herons provide a thrilling spectacle for bird lovers anywhere, but dozens of great blue herons all together—the sight is breathtaking! Nevertheless, this is the spectacle that overwhelms the newcomer here. The tall, black spruces above the marsh in the centre of the island are their nesting places.

Graceful as these large birds seem when flying or standing in shallow water, their extraordinarily long legs and great bodies appear strangely out of place in the treetops. The onlooker might well assume that these birds, like incredible creatures from a Lewis Carroll or Dr. Seuss book, had drunk a magic potion that had made them too large for their perches. However, like all enchanted beings, they seem invulnerable to commonplace rules—and so they do not fall. Summer after summer, these great and mysterious creatures return to preside over their tranquil island to live at peace with the other creatures, and perhaps even to commune with the ghosts of former islanders—French, English, and Aboriginal.

RICHIBUCTO
HARBOR
T.F
TIDAL
FLATS
INDIAN ISLAND
Passe de l'île
SOCK POINT
POINTE DES RICHARD
← TO RICHIBUCTO
CREEK
BEDEC
BAIE
DU
VILLAGE
PIROQUE
GASPEREAU

Indian Island

The Indian does well to remain Indian.

—Henry David Thoreau

This island, the birthplace of the Richibuctos, is ours. It has been paid for by the lives of our ancestors who all lie buried here. Our future lies in preserving our past.

—Chief Peter Barlow

Every major bay and stretch of Maritime coastline was once dominated by islands sacred to our native people. Each of these islands was a gathering place where the Mi'kmaq, the Maliseet, or the Passamaquoddy people of the region held special ceremonies and conducted burial rites. Islands set aside for these purposes tended to be small—rarely more than three kilometres long—oval in shape, and extraordinarily lovely. The site was important since it was here that every member of the tribe would take his final rest. For generations, the first people of this land slept undisturbed beneath the tall trees of their sacred islands, lulled by the music of the nearby sea, free to commune with the spirits of nature that had ruled their lives.

The coming of white settlers changed all this. They often found the strategically located offshore islands best for their purposes as well—purposes generally at odds with those of the Aboriginal people. What happened to "Indian" Island in the Bay of Fundy (Charlotte County)—one of New Brunswick's eight "Indian" islands—in its first hundred years of white settlement was, in many ways, typical of the fate of other Aboriginal islands in the Maritimes.

In 1760 James Chaffey, an energetic and ambitious Englishman, visited the tiny (kilometre-long) island. This treed and pristine place, which the Aboriginal peoples called "little Jeganagoose," seemed just the spot he had been looking for. It was central, fertile and beautiful—from Chaffey's point of view the ideal headquarters for a fur-trading venture. Chaffey wasted little time getting started. In short order they had built a house on the island, which is often referred to as the first on Passamaquoddy Bay despite the fact that an Acadian family had built a house on the island as early as 1704. Chaffey's house, however, was on a grander scale and was to become the centre of his trading empire. From the beginning Chaffey asserted his strong will over the Aboriginals. The Passamaquoddy people watched with awe the takeover of their territory by this dynamic and incomprehensible stranger.

Chaffey prospered, and because of him Indian Island grew to be a trading centre with as many as twelve schooners at one time lying at anchor offshore, waiting to load or unload cargo. Gradually the island drew other businessmen. One of these was a man called Goldsmith, whose business involved boiling sea water to extract salt. In fueling his fires for this enterprise, Goldsmith managed to denude the island of all its magnificent trees.

Now bare and bleak, the once-lovely and sacred island had been defiled. No longer was it a suitable resting place for "The People," living or dead. Like their fathers before them, the Aboriginal people, puzzled by the influx of white settlers to the neighbouring islands where they had camped all their lives, gradually withdrew into unfamiliar territory where many of them perished quietly. On a government

Chief Barlow of Indian Island

mission to assess the potential of the Grand Manan fisheries in 1850, Moses Perley noted a few Aboriginal people camped on the rugged western coast of Grand Manan and fishing just offshore. After that there seems to be no mention of the island's Passamaquoddy people. The Fundy fogs seem to have engulfed them forever.

For the newcomers to the Fundy isles, "Indian" Island continued to be the logical central meeting place. Just like the Aboriginal people, the settlers congregated here to enact rites (albeit, rites incomprehensible to the native people): voting on election day, celebrating militia day, and exchanging songs, yarns, and drinks. Although the new islanders continued to refer to their meeting place as "Indian Island," all that remained of the centuries-old Aboriginal occupation were a few fading patterns beaten into the earth.

Of the "Indian" islands still recorded on contemporary maps of the Maritime provinces, the one that most closely resembles descriptions of ideal Aboriginal islands in their original, unspoiled state is Indian Island in Richibucto Harbour. This island, sometimes referred to as Richibucto Island, has from time immemorial been sacred to the Richibuctos, a dominant band of Mi'kmaq or saltwater Aboriginal people whose territory once extended along the Atlantic coast from the Gulf of Saint Lawrence to the present city of Boston. The Richibuctos had a village at the southern tip of this island both before and after the French "discovered" the island in the seventeenth century. Although all traces of this village have long since vanished, the ceremonial markings on the earth near the original settlement are still visible.

Despite the Aboriginal peoples' continuous presence on or near this site over the centuries, the ownership of the island has been disputed since the seventeenth century, when small portions of it were meted out to various French grantees. Yet even government officials have been troubled by the problem of ownership here, and have been aware of the island's importance to the Mi'kmaq. This awareness has been underlined by the observations of the perspicacious Moses Perley in his reports on Indian affairs in the 1840s. In a communiqué composed for his superior, dated October 21, 1847, Mr. Perley remarked: "The Indians are stated to have a great fondness for Richibucto Island where they have held their annual festival on Saint Anne's day (26 July)." He recommended a quick and final solution to the dispute over the island, whereby the Aboriginal people would acquire clear title to it. Throughout his reports, Mr. Perley's admiration for the Richibuctos and their quality of life is clearly evinced through such comments as: "The Richibuctos are athletic men and noted for their industry."

Mr. Perley's recommendations about Indian Island—as about a number of other matters connected with Aboriginal affairs in New Brunswick—met with official displeasure. For instance, in July 1848 (less than a year after his report on Indian Island), he received a reprimand from a government official, J. S. Saunders, part of which read: "I am directed to inform you that your interference with Indian Affairs when you have not direct authority from the government is calculated to produce dissatisfaction among many of the Indians and to embarrass the Government in their endeavors to carry out the wishes of the Legislature."

Shortly after this missive was sent, Mr. Perley was directed to turn his attention from the Aboriginal peoples to the state of the fisheries on Grand Manan. Yet, in retrospect, the wisdom of Moses Perley's recommendations concerning the Aboriginal people of New Brunswick seems obvious. As W. O. Hamilton and W. A. Spray, professors of the University of New Brunswick's inaugurated Indian Studies Program, remarked: "Had Moses Perley's generally enlightened recommendations been followed in the 1840s, perhaps the New Brunswick Indians would have had less cause for dissatisfaction with the century that followed."

This disputed island, the Richibuctos' traditional rallying place, is densely wooded. Like Shediac Island, it lies low in the water and is essentially an overgrown sandbar barely three kilometres long, with tidal flats occupying much of the outer side of the island. However, unlike Shediac Island, Indian Island is protected from blustery weather in the Northumberland Strait by a long sand dune protruding from the mainland. The inner side of the island runs parallel to the mainland point of land on which the Richibucto people of the Indian Island band have their permanent dwellings. The channel between mainland and island is so narrow here that the distance separating the two seems often little more than a stone's throw. This channel is only just wide and deep enough to deter the casual visitor from crossing to the island to inspect the enchanting wigwams, which stand at widely spaced intervals along the shore just at the entrance to the woods.

This is likely just as well since, from time to time, intruders who have ventured into the Aboriginal peoples' sacred preserves with the persistent intention of violating them in some way have been visited with afflictions as strange and terrible as any recorded in Homeric legends. One would almost think that some powerful unseen force guards this tiny island and its surrounding waters on whose bounty the Aboriginal people have always depended.

One of those who seems to have incurred the wrath of the island spirits was, according to the story, a man who rented a light plane so that he could fly over the island and take aerial photographs with a view to taking the island from the Aboriginal peoples. Since the day was clear and sunny, he apparently flew back and forth over the island many times filming it from every angle through the most sophisticated lenses. However, we are told that he never saw the pictures he took. Shortly after his flight, his sight began to fail, and before the pictures had been developed, he was absolutely blind.

Another story of seeming retribution concerns a man who crossed the river to dig clams on Aboriginal land. He was asked to leave because there were barely enough clams for the Aboriginal people themselves, who had been digging them sparingly for some time so as not to deplete this favourite resource. "Oh," the intruder is supposed to have replied when asked to stop digging. "I can't. I have to dig here since there are no clams left on my side of the river." With the conversation closed in this manner, he continued to dig, filling bucket after bucket and loading them into his boat. The next day he returned and did the same. The Mi'kmaq were heartsick, but felt there was nothing they could do. Force was out of the question; their chief would not permit it.

The third day they could not bear to return to watch the greedy intruder dig, so they were not among the first to learn that his boat had apparently capsized and that he had drowned. Still, with his death the Mi'kmaq were not free of intruders, for within a week of the first clam digger's death, a second digger had appeared. This man, the son of the first, was no more easily deterred than his father. After finding him just as impossible to convince, the Mi'kmaq again withdrew sadly. However, the spirits of the waters must have been strongly offended, for within several days the son was found dead in his bed, apparently of a heart attack.

After hearing such tales, one might well wonder about the powers and attitudes of the man who controlled Indian Island. A first insight into the chief's character—apart from glowing newspaper reports, both local and international—was the comment of one of his neighbours, an Acadian fisherman on the wharf at Richibucto: "The chief, Peter Barlow, is a very good fellow.... He keeps everyone out there in line.... In fact," he added with a twinkle in his eye, "it's too bad we don't have someone like him right here in Richibucto!"

Peter Barlow was a man whose response to life remained enthusiastic and positive. He clearly enjoyed his position as chief, despite the fact that it was, by his own admission, a 24-hour, 365-day-a-year undertaking. He must have served his people well, as he was elected as hereditary chief for forty-eight years (since he was fifteen). He indeed held on to his position longer than any of the other nearly five hundred contemporary Canadian Aboriginal chiefs.

Chief Barlow and his wife lived on an idyllic tract of land within sight of the sea and his beloved island. On his home ground he was surrounded by Barlows, as the plaque at the entrance to Indian Island reservation proclaimed:

> YOU ARE NOW ENTERING INDIAN ISLAND
> THE CHIEF OF INDIAN ISLAND IS A BARLOW
> THE FIRST COUNCILLOR IS A BARLOW
> THE SECOND COUNCILLOR IS A BARLOW
> THE MANAGER IS A BARLOW
> THE COM. DAV. IS A BARLOW
> PLEASE DRIVE CAREFULLY BECAUSE THE PERSON
> YOU WILL HIT WILL BE A BARLOW

Like the above, all signs on the reservation are in English, as are written communications with the outside world. However, in some households—the chief's for instance—the Mi'kmaq language was often spoken. Yet, some of the Barlows' grandchildren who attended English schools do not understand more than a handful of Mi'kmaq words. This is a pity, since linguistic authorities of the past have commented enthusiastically on the poetic and musical qualities of the Mi'kmaq language (much as Grey Owl rhapsodized about the expressive possibilities of the Ojibway tongue).

Reverend Silas Rand, an accomplished Nova Scotia-born linguist who lived for forty years among the Mi'kmaq, said of their language: "Do you ask which is my favourite language? Mi'kmaq. Why? Because it is one of the most marvellous of all languages, ancient or modern—marvellous in its construction, in its regularity, in its fulness." [1] He systematically cited comparisons between the Mi'kmaq language and Greek and Latin. More than three hundred of the Mi'kmaq words he lists have their classical counterparts. Some of the most obvious are:

Mi'kmaq	English	Greek
Padoos	boy	Paidos
Come	a harbour	Come
Ekai	I am going	ko
Oolk	boat	Olkas
Keloos	good	Kalos
Oonuks	wings	Onux

Mi'kmaq	English	Latin
Ait	he says	Ait
Jun	child	Junenis
Agwith	water	Aqua
Qweentum	I ask	Quaero

Face to face with such linguistic similarities, what is one to think? What contact could the Mi'kmaq have had with Latin and Greek long before recorded history? What never-recorded voyages to our shores could have taken place before those of Cartier, Champlain, and Denys to make such linguistic similarities possible? Whereas one is stranded high and dry without clues about the apparent linguistic impact of the Greeks and Romans on the Mi'kmaq, the influence of the French on the language poses no such mysteries.

The appearance of the French explorers bearing new implements and ideas simply made the coining of additional words necessary for the Aboriginal people. For instance, after spotting (perhaps at one of Denys' more permanent encampments) a European type of dwelling—a hitherto unheard of structure with four straight walls set upright to form a square or a rectangle and topped with a roof—they were clearly at a loss to describe this foreign architecture precisely by using their word for their own homes, "wigwam." Consequently, they came up with the more explicit descriptive word, "Wenjeegwom" (French wigwam).

By the time the Mi'kmaq learned that not only the French, but the English and other "foreign" peoples lived in the unusually shaped dwellings as well, the word was too firmly entrenched by usage to be altered. Thus, ironically, today even their own homes fall into the category of French wigwams. Although even thirty-five years ago one could perhaps have assumed that the Mi'kmaq language and culture would soon be extinct, there are signs of a revival of interest and pride in Mi'kmaq culture. This is strongly manifest in the names of some of the chief's grandchildren. They have indigenous Mi'kmaq names, "Gooo" (forest), "Nibooktuk" (pine), "Oonig" (fog), and the youngest was named after the legendary American chief, Geronimo.

Another favourite and symbolic name, which the chief has given another of his grandchildren (as well as a road on the Indian Island Reservation), is "Wilder." This name was dear to him partly because of the gratitude he felt he owed Dr. Wilder Penfield, the great Canadian neurosurgeon whom he credits with saving the life of his son, Second Peter; and partly because of the therapeutic wilderness images that this word evokes. After all, the chief obviously subscribed to the ancient Aboriginal beliefs, which are best

verbalized in Thoreau's epigrammatic statements, such as "In Wildness is the preservation of the World," and "From the wilderness come the barks and tonics which heal mankind." And if Chief Peter Barlow and his band failed to get back their island, what could happen to it? After reading an advertisement for souvenir centimetres of another Maritime Indian Island, one cannot help but fear the same obscene fate might await the island birthplace of the Richibuctos.

In fulfilling such traditional functions as naming his progeny and the places in his domain, and in his role as representative and figurehead of his people, the chief was rather like a reigning monarch. Like any king, he sat periodically for his portrait, which was looked forward to and acclaimed by both his entourage and his outside admirers. However, to continue to hold office, the modern chief must cope with problems that his predecessors must never have imagined in their worst nightmares: the increasing demands of women for recognition, the culturally destructive intrusions of the media, encroachments on wild land, busing children to schools where teachers do not understand traditional ways or values, depletion and pollution of traditional fishing grounds—in a nutshell, the problems which beset most of the world's leaders in our times. Chief Barlow dealt with these and other issues as best he could—often successfully—with the easygoing good temper he shared with many other Maritimers.

As for intrusions on Indian Island, when they seemed so pressing in the seventies, in a grand public gesture in 1975, the chief crossed formally to the island to call attention to past injustices and government lethargy concerning First Nations' rights to the island. Chief Barlow underlined the significance of Indian Island for the ongoing life of his people: "This island, the birthplace of the Richibuctos, is ours. It has been paid for by the lives of our ancestors who lie buried here. Our future lies in preserving and building on our past."

Notes on Indian Island

1. Rand, Silas T., "Introduction." *Legends of the Micmacs*, p. XVll.

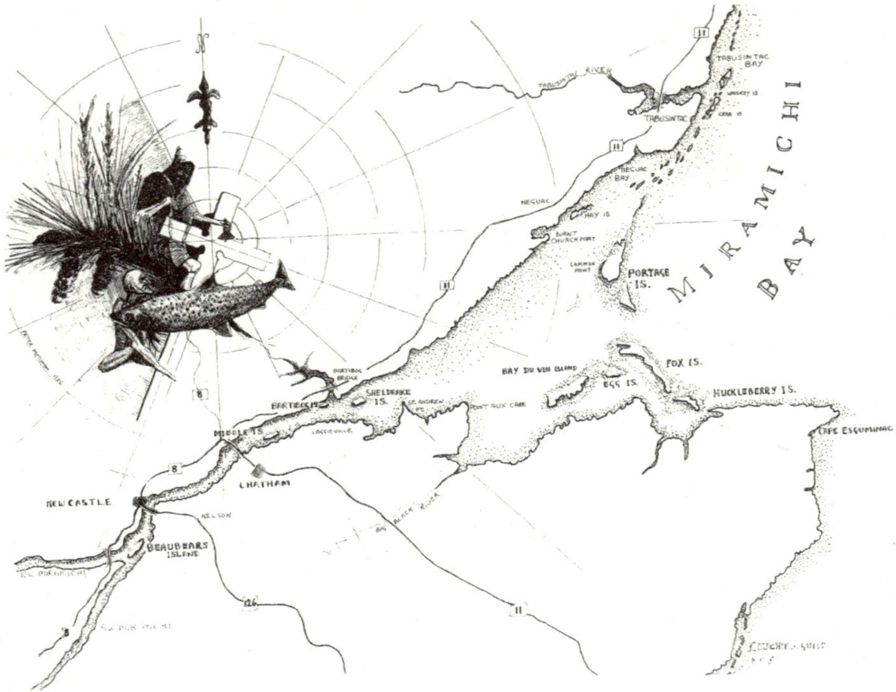

Islands of Miramichi Bay

Islands of the Miramichi

The Salmon put up here were all taken around the island, and were purchased of the fishermen, by this establishment at 3 d. per pound, fresh caught, with a discount of ten per cent for cleaning, which was said to be equal to £3 5s per barrel. The Lobsters were chiefly caught by the French inhabitants of the neighbouring Neguac Villages, from whom they were purchased at 2s 5d currency, per hundred. They were very plentiful the past season, especially at Black Lands and Tabusintac gully; and as proof of the ease with which they were taken, it was mentioned that one Frenchman [Victor Savoy] had, unassisted, caught 1,200 Lobsters in part of one day. There were from twenty to twenty-five men employed at the preserving establishment [on Portage Island] during the season.

—MOSES PERLEY (1849)

I went to the Island [Middle] to nurse my brother Dr. Vondy who was ill with the fever. He was unconscious when I arrived. We were at the South West part of the Island. I still have a memorandum made at the time in my possession. It is this

June 3d 1847. The bark "Loostock" [sic] arrived
Left Liverpool with 467 passengers
117 died on the passage

I do not know how many died afterwards.... The "Loostock's" passengers were in fish sheds on the North East corner of the Island. The vessel was overcrowded and the passengers were crowded down in the hold like cattle. There was not one person entirely well in all that number when they landed. I was not allowed to leave the Island for 21 days. The passengers of the "Richard White" (Barquentine I think) were put in tents over near ours on the S. W. part of the Island. I do not know what disease they had. Drs. Key & Thompson attended the sick. There was much mismanagement and dishonesty charged against those in attendance. The patients were cheated and badly fed and cared for. They were a superior class of persons and had considerable money on them all in gold. When they sent to town for the smallest article they got no change back.

—Mrs. Chas Hawbolt

A Brief Sketch of the River Islands

In few other places along Canada's Atlantic coast have the paradoxes of island life been apparent for so long a time as on the islands of the Miramichi. Here, both paradisal and purgatorial twists of fate have been so commonplace that the fabric of life has been dominated by white-and-black colouring, all but obscuring subtler and more variegated shades. Such a starkly contrasting and arresting pattern has been long in the making. It was, indeed, begun in Aboriginal times. The Mi'kmaq, like the white men who succeeded them, greatly valued the islands in this mighty river. Each one, from the sprawling sand islands at the immense mouth, to the small pine-covered oval some forty-eight kilometres upriver at the forks, contributed to their well-being.

On the shores of the islands at the mouth—now called Portage, Fox, and Bay du Vin—they could dig the finest clams, mussels, and oysters. Just offshore, even in knee-deep water, the most casual wader could pluck lobsters from the bottom, where, in the clear water, they were at once visible; and, a little further out, and still without much effort, anyone could take quantities of herring, mackerel, and cod by net or line. In the channels between the islands, Mi'kmaq and whites alike intercepted salmon on their way upriver—spearing them by torchlight or netting them in broad daylight. On the sea-side of Portage, wallowing sea cows or walrus gave themselves up to the hunter without a struggle; and in the lee of this island, the grey seals were not much warier. They bobbed around and frolicked with one another in sight of their human observers, as trustingly and as enthusiastically as the Labrador dogs they resemble. Geese, both Canada and Brant, stopped regularly to feed on the islands' salt marshes, and sea ducks floated complacently among carpets of seaweed.

Sheldrake Island, almost halfway between the river's entrance and the forks, was likewise a good place to fish, to hunt wild geese and ducks, and to gather succulent samphire greens—as was Middle Island, still farther upriver. Beaubears, at the forks, was always regarded as an exceptionally fine place to camp because of the excellent salmon fishing on both sides of the island and the constant supply of fresh water on the northern side.

These river islands (with the exception of Portage and Fox, which were too exposed to support more than scrub birches and spruces) produced fine stands of pine and oak comparable to those along both shores. Indeed, at the end of the eighteenth century, the pines of Bay du Vin and Beaubears Islands, like their counterparts on the adjoining mainland, were such exceptional specimens that Sir John Wentworth, when he first visited the Miramichi in the autumn of 1788 in search of timber for His Majesty's navy, wrote the following commentary to Edward Winslow: "I have found on this river the best Mast timber in British America...the pine timber for size, length and soundness exceeds any I ever saw in New England." What was particularly remarkable about these trees is that they had achieved their spectacular growth since the fire that had devastated the Miramichi forests just over one hundred years earlier (in 1680, according to Father LeClercq's reckoning).

So bountiful indeed had nature been in endowing this river and its islands that most early visitors assumed that the resources they found here were inexhaustible. Some even complained of an overabundance of good things. One of those who found fault with the river's excessive productivity was Nicolas Denys, when he camped for eight days at the forks of the river opposite the island which the Mi'kmaq called "Quomeneegook" (pine island) or "Niktawak" (at the river's forks)—and which in the latter part of the eighteenth century was to be known as Boishébert's (and from that Beaubears). He remarked: "So large a quantity of them [salmon] enters into this river that at night one is unable to sleep, so great is the noise they make in falling upon the water after having thrown or darted themselves into the air." Even in winter the river provided sustenance: eels by the thousands were taken on lines, and nets full of smelts could be dipped out through holes in the ice for as long as the fishermen could move their arms.

Paradoxically, many tragic episodes in the story of the Miramichi and its islands have arisen from this very abundance of resources—and from the attention, which a succession of literate visitors, both French and English, drew to these treasures. If the Mi'kmaq had been a committed caretaker of the wealth he found around him on this river, others had no such commitment.

The European, sailing into this strange new world and taking stock of its potential, found here a stupendous inventory, which he was convinced he could sell off quickly at bargain prices to a "sophisticated" society that had all but run out of its own natural resources and craved the products of this pristine wilderness. He saw himself as the ideal proprietor, an energetic entrepreneur who—in contrast to the naive and un-materialistic "savage," as he called the Aboriginal—knew how to exploit profitably such a tempting and potentially affluent environment.

Among the earliest literate explorers of this region was Jacques Cartier. In 1535 he reconnoitered the islands at the river's mouth and noted the abundance of life there. His observations were sufficiently interesting to encourage some of his countrymen to set up summer fishing stations on the estuary islands to exploit the teeming waters. Thus began the enthusiastic harvesting of the Miramichi's marine life under French auspices, which lasted for more than two hundred years—indeed until 1760, by which time nearly all of the French dwellers had been expelled from their homes along the river and on its islands by a series of English incursions.

The Anglophone newcomers promptly took up where their predecessors had left off. Although the new proprietors were at first primarily interested in the fisheries, it was not long before they turned their attentions to harvesting the timber that stood so tall upon the river's banks and islands. And so it happened that the islands that were most richly forested—Bay du Vin and Beaubears—became the sites for flourishing mills and shipbuilding. All this activity eventually lured large numbers of immigrants—particularly during the 1830s and 1840s—who brought with them the current European and Asiatic scourges: cholera, typhus, dysentery, and smallpox. As a consequence of near epidemics of these often fatal diseases, two of the river islands, Middle and Sheldrake, were temporarily set aside as repositories for the unhappy victims of these plagues. However, sometimes these islands were used merely as clearing stations for well immigrants, obliged by British law to observe a three- to five-year quarantine period while their ailing shipmates were confined on-board. Sheldrake Island in the 1840s had the added distinction of serving as a detention place for the sufferers from one of the most dreaded maladies of all time—leprosy.

The tragedies that occurred on these quarantine islands in the nineteenth century had had their counterparts in the preceding century further upriver on Beaubears Island. Here, after the dispersion of the Acadians from elsewhere in old Acadie—an exercise which was begun in 1755—a large and desperate band of dispossessed men, women, and children congregated under the leadership of the resistance fighter Charles Deschamps de Boishébert; and here, quietly, over a five-year period, many of them starved to death or succumbed to the various diseases that preyed upon chilled and poorly nourished bodies.

Yet even the islands that have borne witness to the greatest suffering, those on which the grimmest human tragedies have been enacted, have known long idyllic periods. The nightmarish episodes have often been only brief interludes in the otherwise happy and tranquil unfolding of life that prevailed both before and after. Nevertheless, as a setting for displaying the whole range of human possibilities—the extremes of good and evil, and much in between—the Miramichi islands have rarely been matched. They have presented both residents and onlookers with memorable foretastes of heaven and hell.

Portage and Beaubears Islands

When the French fishermen first arrived at the mouth of the Miramichi, the large island near the middle of the entrance was called "Molaweeswayadik" (brant geese, their occurrence place), indicating the importance the Mi'kmaq attached to wild fowl on this island. Gradually, as the foreign fishermen returned to the river year after year and periodically crossed from the islands to north and south banks, the island in the middle of the estuary—the logical stopping place on the journey across—became known as Portage (and sometimes Potage). Meanwhile, the large island nearest the south shore, which the Mi'kmaq had called "Hiktnkook," the French renamed "Baie des Vents" because of the powerful winds that sweep across the water at this place. Upon the English takeover—after 1760—the island and a settlement on the mainland became known as Bay du Vin, presumably the result of the newcomers' inadequate understanding of their predecessors' language.

It was on Portage Island, according to Moses Perley, that the first year-round trading and fishing settlement was established on the Miramichi, and with it the first determined assault on the region's natural resources: "From 1643 to 1647 Jean Jacques Enaud, a native of the Basque Provinces of France, had had an establishment on this Island for taking the Morse or Walrus, and for prosecuting the Fur Trade and Fisheries." Enaud's domains at this time seem also to have included Baie des Vents (Bay du Vin) Island, since both W. H. Davidson and W. F. Ganong have asserted that Enaud had an establishment on this island about 1644 for "collecting fur and peltries." Nevertheless, Enaud was probably not totally responsible for the annihilation of the sea cows off Portage Island; Richard Denys was associated with their demise as well. Ganong has stated: "Tradition also places an early French establishment of Denys at Portage Island, used in hunting sea-cow or walrus." This appears to be a reasonable assumption in view of the fact that one of Denys' settlements is thought to have been at Burnt Church, just across from this island.

Mainland settlements and their attendant fisheries soon followed the pioneering island establishments. In *Historic Sites in the Province of New Brunswick,* Ganong has stated that "Creuxius map of 1660 marks a settlement on the south side of the bay. Another early settlement was near what is still called French River Point, just off the eastern end of Bay du Vin Island." Ganong has added: "The local tradition, as given me by a resident, is that this village pursued the dog-fish fishery for the sake of the skins, which commanded a good price in France, where they were also used for polishing purposes." Cooney, too, has remarked that there was a considerable settlement, complete with chapel, established perhaps as early as 1672 or 1673 directly across from Bay du Vin Island at the place now known as Gardiner's Point. Like the other communities in the vicinity, it depended on a hunting- and fishing-based economy.

Given so much concentrated and blood-thirsty activity at the river's mouth, it is not surprising that by the end of the seventeenth century the walrus or sea-cow—"morse" as the French called this peaceable and vulnerable creature—had been exterminated from the vicinity of Portage Island. After this, the Miramichi residents and other hunters travelled in search of these amiable sea mammals to Miscou, an island where the ecology closely resembles that of Portage. The last of these trips seems to have taken place during the years 1758 to 1760, when the Acadian survivors on the Miramichi (particularly those at Beaubears Island) hunted the few surviving Miscou sea cows for their meat as well as for the fatty substitute for butter that could be extracted from them. By this time the sea cows were as endangered as the Acadians themselves, since most of them had been massacred three or four decades earlier by organized foreign hunting parties.

Not all the early fishermen, hunters, and traders, then, were concentrated at the river's mouth. Nicolas Denys was perhaps the first—and certainly the most powerful and well known—of those who ventured upriver. He must have told his son Richard about the possibilities of the land and river at the forks, and been instrumental in securing him a seigneury there. This grant included "Quomeneegook" (or "Niktawak"), the island situated in mid-river (later known as Beaubears).

By 1672 Richard Denys was established opposite this island. At this mainland trading post he maintained a small fort until, in 1686, he sold the entire seigneury to the Récollet Fathers of Quebec for a mission headquarters. Since the church did not ratify the agreement, the settlement seems to have done little more than hold its own until 1690, when a privateer from Boston destroyed it and took one of its residents, Michel de Gré, as a captive to Boston.

All this goes to show that the land and river in the vicinity of the forks were not exactly unknown and unexplored at the time the Acadians sought refuge on the island there. Ganong has suggested, moreover, that shortly before 1750 some Acadians had reestablished themselves near the forks, almost certainly on the point (now known as Wilson's Point) and very likely on the island as well. Thus it seems that one of the reasons for choosing this location as a refuge after the dispersion was that the already established Francophone settlers would be able to aid the refugees. Added to this, of course, were assurances that the farther upriver they went the less vulnerable they were likely to be to the naval incursions of their enemies. Above all, they were drawn by descriptions of the excellent fisheries on the Miramichi, word of which had circulated as far as Quebec and back again to Acadie.

The French authorities in Quebec had not only recommended that the Acadians retire to the Miramichi for this reason, but stipulated that they would find supplying a camp on the Gulf of Saint Lawrence a much easier proposition than provisioning any such establishment on the Fundy coast or its tidal rivers. Indeed, so auspicious seemed the prospects for the island camp which Charles Deschamps de Boishébert was to lead his people to that it became known as "le camp d'espérance" (the camp of hope).

The location of the camp on this island at the forks proved not to be the real problem. The difficulties that materialized here and on the adjoining point were those that plague nearly all such temporary emergency establishments anywhere, anytime—too many underfed and scantily clothed people set down too hastily in too little space. No island a kilometre long and kilometre and a half wide can provide enough living space for 3,500 people who must subsist on the resources of the land and surrounding water. In a northern climate, the difficulties are obviously compounded by the severity of the weather and the consequent need for substantial houses, nourishing provisions (which can generally only be harvested in the summer), and warm clothing.

Thus, despite the large numbers of fish in the river, game in the woods, trees for houses and firewood, there was not enough of any of these essentials to feed, shelter, and clothe all the men, women, and children who had arrived destitute. The refugees were, moreover, already in a weakened condition after having walked through the woods from the lands bordering the Petitcodiac and St. John Rivers. Even the Mi'kmaq, adept as they were at surviving under difficult conditions, never attempted to sustain so many on so little land.

After several months of privation on their island, most of the Miramichi refugees, because of their numbers and their initially poor condition, were in worse shape than Champlain's men had been after a comparable sojourn on Dochet Island—despite the fact that Beaubears Island was better situated than Dochet, with better land, more and larger trees, and, above all, fresh water. Moreover, although both encampments had turned out to be virtual prisons during the freeze-up when the river ice was unsafe for travel, and both groups of people were particularly susceptible to scurvy because of insufficient and incorrect food, the survivors of Champlain's expedition to the Sainte-Croix could at least look forward to returning in the spring to more hopeful prospects, whereas Boishébert's refugees had no such option. His Acadians had to stay on their island: there was nowhere else to go.

With spring, the beleaguered Acadians could only look forward to having their numbers augmented by other refugees, even thinner and paler than themselves, who would filter in from the woods. Increasingly, too, their hopes of receiving supplies from

CHARLES DES CHAMPS de BOISHÉBERT

Quebec vanished. They realized that a skipper unaccustomed to negotiating the tricky entrance to the river—where even those who were familiar with the passages between the islands had been known to run aground on the sandbars—was unlikely to manoeuvre his ship unobserved past the British blockade here. In addition, they had heard rumors from the woodsmen who travelled regularly between the Miramichi settlement and Quebec that corrupt officials of that town had already bartered away to Aboriginal traders most of the food intended for the Miramichi camp.

Some years later Boishébert's Parisian defence lawyer, piecing together the state of affairs at the refugee camp in 1757 from his client's recollections, produced the following description:

> At the beginning of January, 1757, he [Boishébert] was obliged to reduce the ration of his little garrison to a half-livre [half-pound] of bread a day. Acadians and Mi'kmaq being treated alike. Presently there was no bread. They had kept the hides of the cattle which had been eaten the previous year and they had to nourish themselves on these, as well as on seal oil, of which a small supply was left. This last pitiful store was soon exhausted. All the children died....

Worse was yet to come. In 1758 a French trading vessel, *l'Indienne,* which had recently carried merchandise from the Levant and had spent some time in the port of Smyrna (now Ismir, Turkey) broke up in a storm off Baie des Vents. Among the wreckage thrown up along the coast was clothing which the inhabitants of the struggling settlement were quick to salvage and share out. Those who obtained garments from this wreck soon began to show signs of a horrible disease that in some ways resembled scurvy, the symptoms of which were all too familiar to the refugees. The new plague was, however, even more dreadful and disfiguring than scurvy. Linked as it was to the clothing from *l'Indienne,* this illness was soon diagnosed as leprosy, a pestilence then endemic in the Near East.

Within a short time, eight hundred people are said to have died from the combined effects of famine and leprosy. We are told that the lepers were confined to one part of Beaubears Island, where their miserable shelters were surrounded by a palisade and scanty supplies of food and water were poked in to them through small apertures. As if all this were not misery and hardship enough, Boishébert returned to the pathetic settlement in 1758, just long enough to recruit men to help him in a last-ditch guerilla effort to rout the attackers of the embattled garrison at Louisbourg. Surprisingly, he found those who were able and willing to follow him in this dangerous and ultimately hopeless mission.

Bearing all these facts in mind, the contemporary assessor of the situation at the Beaubears Island refugee camp is amazed to learn that despite cold, hunger, disease, threatened attacks from British vessels at the river's mouth, and the drain of manpower in guerilla manoeuvres, there were still enough strong, industrious, and motivated men in the settlement to accomplish the seemingly impossible feat of obtaining and sending off six shiploads of furs to France. From this commercial venture they hoped to make enough money to purchase both the provisions and munitions necessary for the survival of their camp.

However, the Acadians' ambitious scheme demonstrated not only their resourcefulness, but their naivety as well. The impediments they foresaw seemed to be all in the form of hostile British and New England vessels. Thus they concentrated on evading the blockades placed across their river as well as those positioned near other strategic locations such as the "Passage de Fronsac" (Strait of Canso), and seemingly gave no thought to how little a harassed, slow moving, and corrupt officialdom in Old France was apt to be sympathetic to their predicament.

The refugees could never have realized how low a priority the French authorities gave to provisioning and defending a beleaguered and impoverished encampment far up a strange wilderness river with an outlandish and unpronounceable Aboriginal name.

Indeed, by 1758, astute Frenchmen were far too busy reassessing France's investment in the great fortress of Louisbourg—now once again threatened by the enemy—to attend to any other affairs relating to Acadie. By this time, writing off the unfortified outposts of their New World Empire seemed the only practical thing to do.

Because of such a turn of events, the refugees at Beaubears Island obtained few of the supplies that they had counted on. Nevertheless, the very remoteness of the camp and the difficulty of navigating far upriver in sailing vessels helped the community on Beaubears Island to hold out longer than either Louisbourg or Quebec. Although invaders destroyed the French settlements along the lower part of the river in 1758, they did not get up as far as the old camp on Beaubears Island until 1760, when at last Commodore Byron captured the outpost and had many of its valiant and pathetic survivors transported.

Yet even after 1760, some Acadians remained near the forks and in other encampments along the Miramichi. Surprisingly, at least some of those who remained seem not to have felt particularly hostile to individual Anglophones they encountered along the river. Indeed, the Acadians' kindness to stragglers who fell into their hands is substantiated by Gamaliel Smethurst's account of his 1761 sojourn among the Miramichi Acadians. Like another stranded wayfarer, Witherspoon, who four years earlier had lived with the Miramichi Acadians under somewhat similar circumstances (and also recorded impressions of his stay among them), Smethurst found his hosts ready to share their meagre supplies with him. From Smethurst's account—together with logs from New England trading vessels during subsequent years—one can surmise that pockets of Acadians eked out an existence along the river until 1764, at which time they were entitled to take up various unoccupied lands in the former "Acadie."

The largest grants along the Miramichi were, however, made to Anglophones, the most extensive—which included the eastern end of Beaubears Island—going to William Davidson and his partner, John Cort. Unfamiliar with their new territory and the methods suitable for managing it, these new entrepreneurs took some time to make their presence felt. Indeed, at first they made so little impact that in the summer of 1767 the Jersey trader Charles Robin, on an exploratory trading venture up the Miramichi, made no mention of them whatsoever; and it was not until May 18, 1772, that Robin referred to Davidson in his journal.

Even before setting out on his first journey up the Miramichi, when he tried to find a pilot or guide, Robin was given the impression that the only inhabitants on the river were Mi'kmaq and Acadian, both living a somewhat hand-to-mouth existence. The proposed pilot actually declined to accompany Robin, stating that "he had never been in Miramichy [*sic*] but with log cannoes [*sic*], as the [French] inhabitants there have no other crafts...." (The Mi'kmaq, as Robin observed elsewhere, used bark canoes.)

Robin's journal shows that when he started up the Miramichi he ran into navigational difficulties. In fact, his description of his trip upriver, his sighting of Portage, Sheldrake, and Middle Islands, together with his view of "le vieux camp" at Beaubears Island, gives the contemporary reader a very clear idea of the dangers and difficulties of negotiating the river in a sailing vessel of any size. This account also gives important insights into the wealth of furs and fish to be harvested at or near the river's junction at this time—a fact underlined by the presence of a New England trading schooner, which Robin found anchored off Beaubears Island.

EXCERPT FROM ROBIN'S JOURNAL OF 1767:

July 29th At daybreak weighed, wind at N. W. fresh breeze. At 8 o'clock quite close to the S. W. point of Isle au Potage, took a reef in the mainsail, squally weather down Mainsail, the foresheet gave way, down our graplin in two fathoms of water for fear of being drove on the shoal of the S. E. point of Isle au Potage... tacked about seven times to get thro' the Channel, between two dangerous shoals of moving sand, got thro' at 2 P.M. in two fathoms of water and sometimes less; squally weather out the reefs, and stand by at 5 P.M., reached L'isle d'en Bas [Sheldrake] at 7 reached L'isle du Millieu [Middle Island]....

July 30th This morning at 6 o'clock weighed, quite calm, took to our oars and rowed up, at 8 o'clock anchored before Le Vieux Camp, which was during the French a considerable post, Mons. de Boisbert was Commander in Chief, when the Accadians [*sic*] were drove out of Accadia [*sic*] those that did not pass on the Island of St. John came here and were maintained by the King of France, but very poorly, many of them died in the winter; in this place there is a Boston Schooner that carries on the Salmon and Bass Fishery four leagues higher up the river; the river is divided by this point in two branches, in the one they ketch the salmon and in the other the Bass.

August 1st This morning fine weather, trading with the French till 2 o'clock P.M. sold goods for Livres; accepted for part of payment 100 quintals of dry codfish which I am to take at Tracadie, on condition one of their best men come down with us and for the rest I was paid in Furrs [*sic*], went ashore and dined with one Mr. Morray, one of the owners of the above-mentioned Schooner....

It is not really surprising that Robin failed to mention William Davidson's establishment on the Miramichi in 1767, or even the following year when he undertook a similar trading expedition up the river. Davidson was slow to establish himself in the business that had first caught his interest—the river fishery—because the methods he began by using were more suitable for the rivers of his native Scotland than the Miramichi. However, once he had adapted to his new environment, he was so thorough in setting nets across the river at the forks that few fish escaped upriver.

By 1785 he had already depleted the stocks of salmon so noticeably that Benjamin Marston, a thoughtful young Harvard graduate assigned to survey the Miramichi in that year, was appalled by the methods Davidson used to conduct his fishery. Marston noted:

> The Salmon Fishery on this River is an object worth the attention of the Government, but unless it is attended to will be ruined by the ignorance and avarice of those concerned in it. It has failed very much this season—no doubt thro' the impolitic methods used to catch the Fish, which is chiefly by set nets, which are so extended from each side as to leave the fish very little room to run, and at Davidson's are extended fairly quite across the river to the utter exclusion of the poor savages above.

This entry from Marston's diary sounds very like an abbreviated version of the report on the Miramichi River fisheries that Moses Perley was to make in mid-nineteenth century. Both Marston and Perley were much in advance of their times in advocating reasonable exploitation combined with adequate conservation of resources. With his usual perspicacious insights and his shrewd sense of historic perspective, Perley pointed out that the wretched excesses observable in his own time were part of a pattern begun hundreds of years earlier. He traced the destructive trend as far back as the mid-seventeenth century, to Enaud's base on Portage Island.

Perley obviously felt that his reminders about the extinction of the walrus through entrepreneurial greed ought to alert his contemporaries to what was likely to happen to still abundant resources such as salmon, cod, lobster, and mackerel if suitable checks were not at once imposed on what he referred to as the "illegal fisheries off Portage Island." But Perley's counsel fell on deaf ears. No one seemed interested in hiring the fisheries officers necessary to enforce existing regulations, and no one wanted to think about what had happened in the past or what was likely to occur in the future. Those who made a living from the sea took all they could: they did not think about maintaining the natural balance for future generations.

It was not surprising that Perley was upset by the fishing practices current in the nineteenth century. Nets for salmon were strung out all across the immense mouth of the Miramichi, from island to island, leaving only the ship channel, between Portage and Fox, free. Upriver were more nets stretching from shore to shore. In these, salmon of all sizes were taken from spring until late fall, however sparse the run and whatever the condition of the fish. Meanwhile, offshore, just beyond Portage and Fox Islands, mainlanders from both shores of the Miramichi were busy hauling in cod, ling, lobster, and mackerel as fast as they could. Competing with them were Americans who, we are told, "all obtained full fares." In the summer of 1849 Perley recorded "thirty-three [local] boats engaged in fishing at this station [Portage Island], averaging three men to each boat." In this same location Perley counted between twenty and thirty large and well-equipped American schooners.

The two groups of fishermen appear to have gotten on amicably since there was more than enough fish for all. As their countrymen had done eighty years earlier when Robin visited the river settlements, the Americans of Perley's time brought along goods for barter. Trading as a sideline had indeed long been the practice of many American captains embarked on fishing and whaling expeditions to locations as far from home as the Arctic and the South Pacific. In Perley's time, the Americans' main trading base was Fox Island. Here they sold nets and other commodities, often taking furs and fish as payment. The Americans who fished and traded at the mouth of the Miramichi took their acquisitions home to New England: the local fishermen had their catches processed on the islands close to their best fishing grounds.

About 1844, Perley tells us, sheds had been built near the north-eastern end of Portage Island "with necessary conveniences for putting up Salmon and Lobster in tin cases hermetically sealed." In the summer of 1848, Perley noted, "there had been twenty-two thousand pounds of Salmon, and four thousand pounds of Lobster, put up in tin cases, of one pound and two pounds each.

The quantity of Lobster put up was much less than usual owing to the prevalence of cholera in the United States, and the consequent want of a market there."

These figures reflect only the catches of salmon and lobster brought into the cannery. Meanwhile, cod, ling, and haddock taken in the vicinity of the island were preserved differently. They were dried on the flakes. Perley reported that, "up to the 18th of August (1849), there had been seven hundred quintals [a quintal was 112 pounds] of Cod, Ling and Haddock, caught and cured at this room, to which a considerable addition was anticipated before the close of the season."

The proprietor of this "room" was a Chatham merchant, William Fraser, and 1849 was, according to Perley, the first year he had occupied the fishing station on Portage Island. At this time Mr. Fraser not only oversaw the drying and canning operation on Portage Island, but also established a fish cannery on Fox Island—a venture that proved so successful that in 1853 the commissioners of the London Exhibition awarded him a medal for the high quality of the fish he had preserved there. The awarding of this medal seems somewhat ironic when one considers that the fish, which had been so meritoriously processed, had been caught in open defiance of the law. Despite the law and Perley's recommendations, fish-processing establishments on these islands continued to thrive until at least 1874, and both islands were used as fishing bases until the mid-twentieth century.

Although Portage Island's rich natural resources drew exploiters for at least four centuries, this island seems never to have been considered a good place for permanent human habitations. Most of its various names—Île au Potage (Jumeau, 1685), Île au Portage (Franquelin, 1686), and Île au Passage (d'Anville 1755)—suggest transience. Île au Potage was apparently a designated stopping place where transients boiled up a meal before crossing the bay. And certainly neither "portage" nor "passage" has connotations of permanence. Today, save for the foundation of a lighthouse on the sheltered side of Portage Island, no sign remains of the human enterprises that flourished here; and Portage Island, like its sister islands at the mouth of the mighty Miramichi, has once more become what it was when it was solely occupied by Aboriginals: the nesting and migratory stopping place of wild waterfowl.

Although the much-acclaimed potential of its fisheries had originally drawn traders, refugees, and settlers to the Miramichi River and its islands, entrepreneurs—particularly after Wentworth's 1788 visit—were not reluctant to turn their attention to the other great resource of this region, the forests. William Davidson, before his premature death at fifty (in 1790), had already begun to exploit the vast reserves of timber that began at his doorstep. After his death, his close friend and business associate—another emigrant Scot, James Fraser—took up enthusiastically and energetically where Davidson had left off.

Fraser chose the beautiful and densely wooded Beaubears Island as the site for his mill, his shipbuilding and fishing empire, as well as a store and his own dwelling. In short order, Fraser, together with his partner, James Thorn, had the largest and most profitable business in the province on this island (which, for a time, became known as Fraser's Island). The various enterprises, with their headquarters on the east end of the island, thrived under Fraser, his heirs, and successors for almost a century; indeed until a fire in 1879 burned the mill and most of the lumber in stock—a setback from which the island businesses never recovered.

Of course, the Miramichi timber trade, and related activities such as shipbuilding, had suffered from fires before, particularly from the great fire of October 1825, which raged through and destroyed one third of the forests of New Brunswick. Islands that escaped the conflagration were consequently much valued for their stands of original timber. Bay du Vin Island, with its mixed pine and oak woods, was one of these. Like the mill on Beaubears, the Bay du Vin establishment was on the east end of the island. It was operated successfully until it too was destroyed by fire—though not until early in the twentieth century, according to local residents on the nearby mainland at Gardiner's Point.

Drawn by all the activity associated with the fisheries and lumbering, European immigrants had poured into the Miramichi region throughout the nineteenth century, and, as with their counterparts who had put into other ports along the Atlantic seaboard, they brought with them all the most dreaded diseases of their time. Predictably, like the residents of other communities both in Europe and in the "New World," Miramichiers were quick to relegate the sick to offshore islands where they believed the contagions would run their courses with minimal danger to themselves.

Middle and Sheldrake Islands

Middle Island was used as a hospital island on and off for nearly one hundred years. Sheldrake Island was hastily purchased in 1832 for use as a quarantine centre. Then-panicky residents of Chatham, after learning of an outbreak of cholera in Europe, thought that they could hold the disease at bay if immigrants arriving on the Miramichi ill with the dread disease were kept as far away as this island. The spur-of-the-moment purchase of Sheldrake Island backfired on the community organizers, however: it was just too far from the town for doctors to travel there regularly to treat patients. And so, year after year, the immigrant ships with their tightly packed and suffering human cargoes were put ashore on Middle Island for periods of from twenty-one to thirty-five days, depending on whatever quarantine restrictions were then applicable under British law, and others—those unable to survive the various diseases and the miserable accommodations—were left forever.

Sheldrake Island was not used regularly as a hospital island until 1844, and even then the sufferers there were not European immigrants suspected of harbouring the germs of typhus, smallpox, cholera, or dysentery, but native-born New Brunswickers smitten mysteriously with leprosy. For their mentors, the remoteness and inaccessibility of Sheldrake Island seemed all at once very much in its favour. No one, it appeared, ever anticipated giving the lepers even the minimal amount of care and attention sometimes accorded the hospitalized immigrants.

To refer to Middle and Sheldrake Islands and their counterparts at other strategic locations along the Maritime coasts as "hospital" islands was generally a mistake in nomenclature. The Miramichi "hospital" islands seem to have had particularly inadequate facilities for the hundreds who were confined there from time to time. Ironically, it was not until early in the twentieth century, when very few new arrivals on the Miramichi needed to be quarantined, that the two small hospitals—of about a dozen-bed capacity each—were remodelled and brought up to a first-rate standard.

This refurbishing was done under the watchful eye of Dr. Baxter, the quarantine physician who first took office at the beginning of March 1907. Up to that time, conditions at the Middle Island facility seem not to have been much better than they had been throughout the nineteenth century. As Dr. Baxter noted in his diary almost two months after his takeover of the facilities: "An inventory was forwarded to Ottawa of all the material at the island and it was a short one as there was simply nothing there and the tube of the Artesian well was frozen 6 feet deep, the house bare and dilapidated and dirty. I obtained permission for repairs and proceeded to carry them out." As his journal indicates, Dr. Baxter became so wrapped up in setting things to rights that he often did the jobs which needed doing himself—fixing a loose board here or applying a new coat of paint there. Bit by bit the island property began to look well groomed. Dr. Baxter had the steward's house raised above the reach of high tides, had bathrooms and running water installed in the hospitals and the steward's house, and had a supply of coal laid in to spare the steward and his family the arduous and often frustrating search for driftwood.

Yet in all the years (1907–19) Dr. Baxter presided over the Middle Island quarantine station, there were few patients in the island hospitals. Year after year his report ended with the statement: "No quarantinable disease"—this verdict, despite the hundreds of vessels and their crews that he inspected. The few patients who were detained on the island "were treated like family, their meals cooked in the house and brought out to the hospitals on trays," as Mona and Catherine, daughters of Mr. McFarlane, the last steward, recalled recently. "Those were the best of times," reminisced Mona (McFarlane) Kingston:

> The fine big house surrounded by lawns, the boat we rowed to school in, the lighthouse keeper who spent so much time with us when we were little and who was so patient and, taught us to tie our shoes, the few patients we looked after and who, as they got better, became our friends, and the doctor we all looked up to. Our parents even called our brother after him, and his nickname was 'Doc' for the rest of his life. Probably no one else in the world ever had such a wonderful childhood as we did on that island.

Indeed, conditions during Dr. Baxter's island reign could hardly have presented a greater contrast with those that had prevailed in the same spot seventy years earlier, when each arrival had seemed inopportune, unexpected, and unwelcome, and when each ship bearing sick passengers had caused an emergency. Then, particularly between 1827 and 1848, shiploads of ailing immigrants arriving on the Miramichi always seemed to catch the river communities off guard. In a bad year the authorities threw up a rough shed or two to house the ill, and when at length the sufferers had either died or recovered, those who had taken charge apparently put the whole tragic problem out of their minds. The hospital sheds were then used as fish houses until the next ailing newcomers arrived.

When Middle Island was first used as a quarantine centre in 1827, the disarray of the Chatham townspeople was understandable.

The great waves of immigration were only just beginning and no one knew what to expect. Besides, the communities along the Miramichi were not in the best position to welcome and provide for newcomers—they could scarcely maintain themselves. They were only just beginning to recover from the devastating fire of 1825. Not only had this fire razed most of the buildings along the river, with the exception of many of those in Chatham and a few others here and there, but, in burning the forests, it had deprived the majority of the residents of their livelihood. Thousands of acres of prime forest had been burned, and even ships anchored in the river had been gutted. Both before and after the fire, hundreds of people had sickened and died due to the unwholesome conditions said to have been brought on by the unprecedented heat and drought, which had distinguished that disastrous year.

So in 1827, when ship after ship arrived with sick immigrants, it is small wonder that the Miramichi residents were ill-prepared to cope with them. They did their best under the circumstances. By the end of May, the river communities were swamped with newcomers—by then fifteen hundred had arrived. Passengers from some of the ships to make port that spring—the two hundred and fifty immigrants aboard the *Foveran,* for instance—were quarantined at Saint Andrew's Point; but, as summer came and the flood of newcomers continued, the Miramichi residents showed their increasing resentment at being exposed to the often fatal diseases that many of the new arrivals brought with them.

With the arrival of the *Elinor* and the *Mountaineer* in July—both carrying large numbers of sick—the community seems to have sprung into action. Virtually overnight, the men got together and threw up a quarantine shed on Middle Island, to which the sick were transferred from the anchored ships. Yet despite the daily deaths, this thirty-bed facility was unable to cope with even half of the seriously ill patients. By the end of July every bed was occupied, and more than thirty equally-sick immigrants were waiting to enter. There had to be another hospital.

At an emergency public meeting on August 1, 1827, the townspeople decided to construct a second building at once. They must have started it the following morning, because by August 4 the frame had been hammered together and transported to the island. By August 5, the roof had been boarded in and the shingling nearly finished. By August 7, the new hospital was ready for occupancy. There had been no time to apply for outside help, although the people of Chatham who had put up the money expected the government to reimburse them.

The next problem year—a minor one compared with 1827—seems to have been 1830, when the *Jane*, out of Cork and bound for Quebec with 170 passengers, was diverted to the Miramichi because of illness aboard. Since the numbers of sick apparently far outnumbered the well, the Middle Island facilities were used for the healthy passengers while the ailing were confined to the ship. Two armed boats patrolled the waters around the island and the ship so that none of the quarantined passengers could swim ashore. Despite such war-like measures, word somehow reached town that the imprisoned immigrants were starving. When Henry Cunard, the immigrant agent, verified these rumors, he and a number of other citizens set about collecting money and buying provisions for the helpless newcomers.

If the situation was appalling for the immigrants, it was not exactly ideal for the nearby Chatham residents either. Middle Island was only just outside town. It had become increasingly clear during the periods when immigrants were incarcerated on the island that even a weak swimmer stood a good chance of getting across the narrow channel to the Chatham shore at low tide. Thus, the townsfolk felt that they were almost certain to be exposed to whatever diseases were current just offshore. Moreover, because the immigrants were confined so close to their town and the government could not be counted on to be responsible, the townspeople found that the burden of housing and feeding hundreds of newcomers kept devolving on them. Understandably, this handful of permanent residents were fed up, convinced that the immigrant detention centre should be moved to a location farther from their town, farther too from any mainland shore, and quite beyond the jurisdiction of any one small community.

Consequently, in 1830, after the departure of the *Jane*, the community was relieved when a Chatham merchant, A. D. Shirreff, offered to buy the island and establish a fishing station there. He could take over the buildings for fish sheds, meaning that the townspeople would no longer be out of pocket for financing the island "hospital." Shirreff's offer came at just the right time. Moreover, since he had formerly had a house on the island and had cleared much of the land around it, his proposal to move back to Middle Island scarcely seemed controversial. Shirreff's island house had indeed burned in 1828, but the still-intact "hospital" buildings would presumably compensate him for that loss.

One can speculate that the town heaved a united sigh of relief when Shirreff turned the hospitals into a large (thirty-seven by fifteen-metre) fish shed and flamboyantly renamed the island "Barataria"—in memory, so he said, of Sancho Panza's Island. There seemed to be little doubt that the fishing establishment was much better planned and managed than the immigrant station ever was. Shirreff ran out seines from the northeast and southeast corners of the island. When these were full and the tide falling, he had horses draw the nets in. The fish (shad and salmon) were then dumped into carts, which were brought to the shed (situated on the bank of the northeast side of the island) where they were removed with special wire shovels into upwards of twenty stationary vats. These tanks, placed in the centre of the building, had a capacity for holding and curing several thousand barrels of fish.

And so it happened that Shirreff conducted his island business tranquilly, until he became indebted to Joseph Cunard and forfeited the island. Cunard was content to use this cleared and cultivated land as a pasture, and the fish shed as a warehouse for storing the commodities necessary for his many business enterprises. And if, as we are told, a horse fell through into the immigrant graves that enriched his grass now and then, Cunard was not distressed. He was too busy with all his money-making schemes to even worry about Shirreff's undisguised animosity.

Although the Chatham townspeople probably never dreamed that the island would ever again be used as a quarantine station, they were due for a surprise. Certainly, Sheldrake Island had been purchased in 1832 and set aside for this purpose, but by 1844, this more remote island was being used as a leper colony. And so, in 1847, when there was another influx of sick colonists, the river inhabitants once more had to face up to the problem they had confronted two decades earlier: What was to be done with hundreds of ailing immigrants?

Many people were all for housing the newcomers in tents or fish sheds alongside the lepers on Sheldrake Island. The lepers, however, objected violently to this plan. Already ostracized and mistreated, they simply could not abide the idea of being exposed daily to new and virulent diseases. And for perhaps the only time during their stay on Sheldrake, the lepers had their way. They were informed that they could keep the island for themselves—at least for the time being. This meant that Middle Island had to be used once more as a quarantine centre. Again the fish sheds reverted to their original function as hospitals, although no work seems to have been done to make them in any way habitable for human beings. When no more immigrants could be crammed into the sheds, the human overflow was contained in tents. This was the situation that existed in late May and early June of the fateful year, 1847.

As the summer of 1847 wore on, life for the residents of Chatham scarcely seemed worth living. Few were able to enjoy the fine weather. Fear of the pestilences—supposedly contained on the threshold of the town—hung over the river community. With every breeze, the townsfolk worried about the spread of infection from the nearby island. Even the river water, many surmised, was apt to circulate typhus or smallpox, dysentery or cholera. Children were forbidden to swim or wade in the river. The usual picnics and clambakes along the shore and on the river islands were cancelled. When some imagined that the fish might be contaminated, the fishermen moved their weirs away from the vicinity of the island.

The whole situation on Middle Island was complicated by the fact that proper facilities and a permanent quarantine centre could not be established there because Joseph Cunard refused to sell the island for this purpose. Consequently, in March 1848, the government ordered the board of health to surrender the upper part of Sheldrake Island to the Justices of the Peace for a quarantine station, noting that this was the purpose for which they had originally purchased the whole island.

It seems unlikely that the government representatives who made this recommendation had ever visited Sheldrake Island or consulted anyone who had, since most of the upper (upriver) half of this island is a swamp, obviously unsuitable as a site for any building. Perhaps, since the proposal was made in winter, everyone conveniently forgot about the swamp. In any case, winter was fixed upon as the appropriate time to re-establish Sheldrake Island as an immigrant quarantine centre. This was the only time that the old "hospitals" (or fish sheds) could be hauled down the ice from Middle Island.

However, arguments between the members of the board of health and the Justices of the Peace as to which group had legitimate claim to Sheldrake Island lasted so long that the spring thaw was almost underway before any action was taken. Then, on April 3, when it seemed that the ice would not hold much longer, a strange procession set off downriver. Leading the parade were several teams of draft horses straining to pull the old sheds whilst trying to keep their footing. To one side, Joseph Cunard, cantankerous and opinionated as usual, cursed and gesticulated, exhorting both the Justices of the Peace (who had fallen in behind him) and the teamsters he had hired to get a move on. Bringing up the rear was a ragged band of men, each carrying a gun over his shoulder and struggling futilely to maintain a military bearing on the slithery surface.

As the horses reached Sheldrake Island and were attempting to haul the old buildings onto the low end (the end closest to Middle Island), a confrontation occurred. Out to meet and oppose the motley band of invaders was an equally bizarre force. A twelve-man contingent—hired by the board of health and backed by the lepers—who continued to be hostile to the idea of being exposed to new diseases, blocked their way. A struggle ensued in which guns were used as clubs. In the resultant battle one magistrate was hit over the head and obliged to withdraw from the action, whereupon the rest of the aggressors picked him up and retreated upriver. Since the board of health battalion had won the day, the lepers continued for another year (until April 18, 1849) to occupy Sheldrake Island. Meanwhile, throughout the summer of 1848, Middle Island had to be used again as the quarantine station.

This odd confrontation over Sheldrake Island, though appearing somewhat comical from a contemporary vantage point, nevertheless had sinister implications. Several of the defenders of Sheldrake were not just engaged for a day to make a point and a show of force: they were part of a full-time guard which patrolled the lepers' island in a not-always-successful attempt to keep the lepers in and everyone else out. Since 1845, the lepers' second year on the island, the hellish terms of their confinement had resembled those of the worst prison imaginable in countries lacking the advantages of Canada.

When one looks into the conditions under which they lived, one is amazed to learn that any of the lepers survived to make the 1849 trip to their new lazaretto at Tracadie. On Sheldrake, running a constant fever, they shivered or sweated—depending on the season. The sheds they inhabited were so badly constructed that in winter snow drifted in and settled upon their thin and filthy clothing, and in summer hordes of mosquitoes flew in from the swamp and settled on their exposed and running sores. No

"Then on April 3, when it seemed that the ice would not hold much longer, a strange procession set off downriver."

wonder the Mass for the dead was said *before* each leper entered the Sheldrake lazaretto. Although the lepers' ordeal in some ways resembled that of the immigrants quarantined for various diseases on Middle and other "hospital" islands, the prospect was different. For the immigrants, release was imminent. They could look forward either to a hasty death or a cure to terminate their incarceration. If they did not fall ill, the period of imprisonment was definite—three or five weeks, depending on the current law. The lepers' sentence was lifelong.

Despite the fact that segregation of lepers (and immigrants too) in island quarantine detention centres was subscribed to throughout the world, nowhere were the circumstances governing incarceration more sordid than on Sheldrake Island. One thinks, for instance, by way of contrast, of the better regulated leper colony on the Hawaiian island of Molokai. Here, on a grassy eight-thousand-acre elevated plain, hospitals, dispensaries, churches, and comfortable, well-ventilated cottages were eventually constructed to spare the inhabitants as much physical and mental suffering as possible. Patients who wished to be treated in their individual cottages were attended there by the resident physician. When possible, administrators made sure that each Molokai resident drew his twenty-one pounds of fresh beef weekly, together with all the taro, flour, bread, rice, tea, sugar, salt, tobacco, and good clothing he needed. If he did not claim all of his allotment, he was reimbursed in cash, which he could spend as he wished. Although Molokai was still not exactly a tropical island paradise because of the dreadful nature of the disease and the fact that, even here, confinement was not voluntary, the Molokai lazaretto nevertheless seems ideal when compared with its Miramichi counterpart.

And yet Sheldrake Island was not, by its own nature, antithetical to comfort. With proper attention, with a dedicated and humane administration, it could have provided just as suitable an environment for a lazaretto as Molokai. There was clearly no need for the handful of New Brunswick lepers to be subjected to so much humiliation and suffering. The island could have supported the fifteen to twenty sufferers interred there in relative comfort. Its woods and beaches, clam flats, and river fishery provided many of the ingredients for a satisfactory way of life. This was proven by the prosperity and contentment of later dwellers on this island. Lawrence Comeau, an elderly resident of Loggieville, recalled the many happy hours he spent on Sheldrake Island during his childhood. His maternal grandparents, the Dukes, ran the lighthouse, fished, and farmed on the northeast side of the island. His uncle farmed, fished, and hunted from his house on the southeast side. As Mr. Comeau recalls, it was a place of plenty. It was the one spot he can remember where there was always more than enough of everything good—wild game, fish, clams, eggs, garden vegetables, and firewood.

Today, a mariner who enters the Miramichi through the channel between Portage and Fox Islands (originally Île passage de Jumeau) must see almost the same sights as those that greeted his predecessors, the late seventeenth- and eighteenth-century fishermen and traders. Save for a few navigational markers and lights on both islands, several camps on Fox Island and the foundation of the former lighthouse on Portage Island, the wave-sculpted sandbars and scrubby vegetation indicate that, once more, there is no permanent human presence here. A century's winds, tides, and new growth of trees have obscured all traces of the nineteenth-century fishing shanties and canneries.

A Contemporary Perspective

The contemporary traveller by sea will also note the fishing boats, still present, are no longer powered by sails, though their holds are filled with lobsters or mackerel (in season) as in centuries past. However, salmon are noticeably absent in their tallies.

After negotiating the entrance channel, the present-day mariner who veers towards the river's south shore skirts the island of Bay du Vin, which is, as in bygone days, covered with mixed spruce, fir, pine, and oak woods. The trees have grown up over the old mill site, which faced seaward, and over most of the cleared pasture and hayfield on the inner, more protected, side of the island. The treasure, said to have been buried centuries ago in the sand dune that joins the east and west portions of the island, has long since either been dug up and taken away or lost among the shifting sandbanks. Nearby mainlanders who have watched the comings and goings on the island opposite for generations reckon that despite all the energetic digging that strangers have engaged in over the years, the treasure is still somewhere on the island. Perhaps, some suggest, it has been protected by the sacred nature of the church ornaments, which local residents maintain constitute a part of the buried valuables.

The herons, which nest inland on this island now have, as one awestruck visitor remarked, "nests so big that a man could stand upright in one of them." The birds seem to think that for the first time in centuries their rights are uncontested here—that the wildlife official who occasionally puts in to their shore has come to protect their rights to this island paradise.

Yet these wild creatures, as other observers have noted, could well have been mistaken in their choice of habitat. Although the island has been set-aside as a provincial park, boatloads of hunters put along its shores in November hunting ducks and geese. Others go there to take potshots at the so-called nuisance species—crows, blackbirds, and cormorants—that cannot legally be shot at any time of the year. Not all of these gunners are careful or knowledgeable enough to distinguish between cormorant and heron, crow and osprey. Unfortunately, it is not possible to keep a close watch on the island and on those who misuse it, as the provincial natural resources personnel are the first to admit.

Middle and Beaubears Islands are now government property too. On both these islands, the contemporary visitor is somehow surprised that only plaques remain to remind him of the thousands of human tragedies that occurred here. Although whenever much earth is moved aside in either place, skulls and bones are turned up, the surface of these islands now yield little evidence of either the purgatorial or paradisal interludes in their histories. For the visitor intent on recapturing the past, Beaubears is the better place. Here he can walk down the middle of the island on a pine-arched road that dates from French times, drink from the pure spring on the north bank—beside which the Mi'kmaq and traders once camped—and inspect the many indentations along the south bank that attest to the shape and number of former buildings.

However, it is the giant pines that make this island so special and memorable. None of the other Miramichi islands, and few mainland locations in the Maritime provinces, now possess a comparable stand of pines. The preservation of these trees and the generally unspoiled state of this island are chiefly due to its former, eccentric owner, Lieutenant-Governor O'Brien, who watched over this piece of much-prized real estate until his death in 1972, when he left it to the province. On his retirement to the family home opposite Beaubears, O'Brien seems to have spent much of his time guarding his island. Summer and winter, local residents relate, he watched suspiciously for intruders from behind the ivy-covered windows of his mainland residence. If he spied a boat putting in to the south shore or a wisp of smoke curling above the trees, he was on the warpath. At once he set about having the intruder evicted and reprimanded.

Yet O'Brien's obsessive protectiveness of this island, his by and large successful struggle to maintain its pristine quality, is not without its ironies. It was, for instance, the O'Briens who sold the lovely pine grove adjoining their own property at Nelson—a stand comparable to the one on Beaubears Island—to mill owners whose first act was to level the woods. Within several years the mill had polluted the entire south shore of Beaubears Island as well as the nearby mainland shoreline. And since there was another plant—a paper mill—on the north bank just above the island, this pristine oval sat in a swirl of toxic wastes. There seems to be little likelihood of residents at the forks on either shore ever again being kept awake, as Denys was, by the schools of leaping fish.

It is difficult to predict the fate of this great river and its lovely islands. So bountifully endowed by nature and so frequently misused by human beings, their destiny still hangs in the balance.

One would hope that there is still time and inclination to restore this environment to some measure of its original purity, where fish and fowl, pines and people can balance their needs.

Notes on The Miramichi Islands

1. Charles Deschamps de Boishébert et de Raffetot was one of the most controversial and picturesque figures ever to enter or emerge from the woods of colonial "Acadie." A descendant of a noble, ancient, and affluent French family whose grandfather had settled on a Quebec seigneury, Boishébert was born in Quebec City on February 7, 1727. At fifteen he entered military service, and at nineteen, a year after the first English conquest of Louisbourg, he was part of a seven hundred-man expedition commanded by his uncle, Jean-Baptiste de Ramezay, to retake Louisbourg. On their way, this expeditionary force stopped at Bay Verte where the Mi'kmaq informed them that a British warship and transport were lying off Île Saint-Jean (Prince Edward Island). Thinking that the English vessels could be destroyed easily, de Ramezay sent his nephew to reconnoitre.

This seems to have been the first of Boishébert's remarkable exploits. He crossed the Straits in a canoe, returned promptly to report on the vulnerability of the enemy, and then took part in a successful surprise attack. Such manoeuvres typified Boishébert's frequent and generally well-managed forays into enemy territory over the next thirteen years—that is, until the siege of Louisbourg in 1758.

Despite his remarkable record up to this time, Boishébert's last Acadian mission—to harass the enemy encamped outside Louisbourg—was such a fiasco that it all but ruined his earlier and seemingly untarnished reputation. Instrumental in dishonouring him were two prominent clerics, the Abbé Manach and the

Abbé Maillard. Both wrote letters to their superiors condemning Boishébert for cowardice, indecisiveness, and profiteering. The Abbé Maillard, whose words were invariably listened to because of his reputation for integrity, wrote several particularly damaging commentaries. They are best summed up in two short sentences: "This officer [Boishébert] was not at all the man we needed. He had the soul of a trader and is too rich to run the risk of gathering laurels on the field of Mars."

Actually, Boishébert's last Acadian mission remains something of a puzzle. A contemporary reader can still share Abbé Maillard's wonder and disappointment that Boishébert showed so little enthusiasm and resourcefulness at Louisbourg. At the same time, in the light of all the evidence (that, for instance, Boishébert had brought with him a mere two hundred or so ill and unshod Mi'kmaq and refugees instead of the 1,200 troops he was thought to have had), the resistance hero still retains much of his lustre and elicits sympathy. Out of the dusty pages of past inquiries, he still emerges with much of his original aura—an Acadian folk hero, the real-life prototype indeed of future fictional resistance heroes, comparable in style and grace to Baroness Orczy's Sir Percy, "the Scarlet Pimpernel," who, during the French Revolution, specialized in snatching innocent victims from the greedy jaws of the guillotine and spiriting them to safety.

Yet despite Boishébert's longstanding reputation as a sort of "Scarlet Pimpernel" of the woods of "Acadie," it is impossible to ignore his temporary disgrace after the final British conquest of Canada. Although he had jumped from his hospital bed to rush to the defence of his native town on the Plains of Abraham, even this action, combined with his heroic rescues of thousands of Acadians, did not save him from imprisonment in the Bastille after the fall of Quebec. His incarceration doubtless resulted from the accusation that Fathers Maillard and Manach had made several years earlier. In any case, Boishébert languished in prisons for fifteen months, accused of the same misdeeds as Bigot and his associates, until he was tried and vindicated. On regaining his freedom, Boishébert retired to the family estate at Raffetot (near Rouen) of which he was seigneur. There, after marrying and then fathering at least one child (a son, Roch Charles Gabriel, 1762) he lived out almost exactly his allotted three score and ten (dying January 9, 1797).

BAIE DES CHALEURS
MISCOU LIGHT
MACGREGORS MAL BAY
MISCOU ISLAND
WINDSORS MAL BAY
MISCOU CENTER
SANDY POINT
MISCOU GULLY
MISCOU HARBOR
BLACK PT.
LITTLE SHIPPEGAN
PORTAGE BAY
SKAIT PT.
SHIPPEGAN ISLAND
POINTE CANOT
113
CAP BATEAU
POKESUDIE ISLAND
LAMÈQUE
SAINTE RAPHAEL SUR-MER
ÎLE LAMÈQUE
SHIPPEGAN
CHIASSON
SHIPPEGAN GULLY
PETIT POKE-MOUCHE GULLY
N

Miscou Island

A wooden lighthouse, reckoned to be perhaps the oldest of the original lighthouses built along the Northumberland Strait, stood sentinel at Miscou Point on this historic and enchanted island that is so strategically situated at the northern end of the straits and the entrance to the Bay of Chaleur. Built in 1856, this lighthouse provided direction for brigs and schooners as well as for their motor-powered successors over the years, although it seems to have been of no use to the mysterious ghost ships, which have been reported burning and foundering off Miscou just as often in the last two centuries as in the two and a half preceding centuries of this island's extraordinary history.

This lighthouse was constructed relatively late in the four hundred years of Miscou's recorded history. Few places in Canada can boast of having records that go back so far into the past (only Cape Sable Island, the LaHave Islands and Scatarie Island are possible island contenders), and even fewer have had such varied and unusual visitors and residents that have contributed to an outstanding island folklore. Cartier and Champlain were among the earliest and most famous visitors; Nicholas Denys and the Jesuit fathers among the first dedicated visionaries; then "Gougou," resident sea monster, perhaps the largest, fiercest, and most exotic monster on record since the Homeric legends.

Despite its remarkable past, contemporary Miscou seems, at first glance, unremarkable. It has neither the lush growth nor the dreamy, pastoral charm of other Northumberland Strait islands

such as Pictou, Port Hood, or Cocagne, nor does it have the stark beauty of some of the rockbound islands off Nova Scotia's eastern, southern, and Fundy coasts. Miscou sits low in the water—indeed it is gradually being inundated—and is ringed with fine, sandy beaches. Almost half the island has always been unfit for human habitation (though a paradise for water birds, hunters, and berry-pickers), because of numerous bogs and ponds. The island's name is, in fact, derived from the Mi'kmaq word "M'susqu," meaning "low or boggy ground." The habitable portion is meadowland or sparsely treed woodland. The trees, however, seem never to have recovered from a fire which, according to Denys' account, ravaged the island in the seventeenth century when a gunner who was getting his powder dry on the island accidentally exploded it while lighting his pipe.

The island houses appear to be comfortable and the churches well maintained and suitable for their congregations, but none of the buildings now standing date back before the nineteenth century. The Mi'kmaq, the fishermen of many nationalities, the traders, the explorers, the missionaries—all those who occupied Miscou before the final resettlement of the island in 1815—disappeared without leaving much evidence of their occupation. Unfortunately, nothing remains of the Jesuit chapel, built during the missionaries' sojourn between 1635 and 1662, save the broken remnants of a hand-made chalice, which was unearthed some years ago near the grass-covered foundation of the ancient church. Neither does evidence of the elaborate seigneury—complete with fruit trees—that Nicholas Denys established for himself in 1652 on Shippagan Island just opposite the mission, remain, except for a basement.

Over the years, visitors' and residents' reactions to Miscou have varied tremendously. About no other island have responses been so disparate. For some, Miscou has seemed one of the most idyllic spots on earth; for others, one of the bleakest. The following remarks, the first by the famous historian, W. F. Ganong (1906); the second by a Jesuit priest (1947), represent both points of view. Even today, these opposing viewpoints have their supporters.

A part of Ganong's eulogy reads:

> Miscou;—ever to me an island of charm. For I find it goodly in clime and fair to see; storied of old and ancient today; strange in form and forever in change; haunt of wild life and home of kind men; our Ultima Thule, great for the student and seeker of rest. (*History of Miscou*)

The Jesuit account begins with the following statement:

> The island of Miskou, about seven leagues [twenty-eight kilometers] around the perimeter, is situated in the Gulf of Saint Lawrence...; the soil is not good, the water is not wholesome, the trees are neither so tall nor so beautiful as on the mainland.... (*Relations des Jésuites*)

The Jesuit's disparaging tone is not difficult to understand when one learns more about the priests' lives on Miscou in the early seventeenth century. Those who were sent to the island, with the exception of Father André Richard and Father de Lyonne, were not robust men. Scholarly and ascetic, dedicated and faithful to their last breaths, they seemed as surely doomed to die, victims of an alien environment, as were their brothers, martyred by the Iroquois in Upper Canada. And from all accounts, the sufferings of Miscou's pioneer clerics were at least as drawn out and horrible as the dramatic tortures undergone by more famous martyrs.

Winter, often harsh on Miscou, was the season to be dreaded. Ice on the gulf brought an end to the comings and goings of the boats which all summer had plied regularly between Quebec and the island. The little settlement, where a bountiful harvest of fish and skins contributed so much to both new- and old-world prosperity, was on its own from November to May. As had happened with earlier settlements on Miscou during Champlain's time, nearly everyone left the island for the winter, so that only a handful remained behind—chiefly to look after fishing gear that would be needed the following summer. Champlain, obviously appalled by the difficulties of life on Miscou in this season, recorded how in the terrible winter of 1627, two metres of snow fell on the island between November and the end of April, wiping out the garrison left there.

The Jesuits did not abandon their flock during the winter, although the difficult environment, together with the attitudes of some of their parishioners, would have made most ordinary men acknowledge defeat and leave after a first encounter. The missionaries were constantly reminded of the fact that their parishioners, particularly their recent Mi'kmaq converts, needed frequent manifestations of benign, divine intervention in practical matters to maintain their faith and that their main focus was, of necessity, simply physical survival.

That the missionaries understood the Mi'kmaq conflict between religious commitment and practical necessity is evident in the following story, which was related by an early Miscou priest. One Christmas Eve, the priests were waiting to celebrate Midnight Mass. Two Mi'kmaq set out for the chapel, but three leagues (twelve kilometres) from their cabin, they encountered bear tracks. Since their houses were already haunted by the spectre of famine, they were tempted to go hunting rather than attend church. New snow was fast covering the tracks, so there was no time to waste. Yet despite the strong temptation and their pressing need, the Mi'kmaq did eventually decide to forego the hunt and fulfill their religious duties first, as they had promised. Later, returning home from church, they again encountered fresh tracks, and this time they followed the tracks and killed the bear, thus providing themselves and their families with food for many days. These Mi'kmaq were, we are told, once more strongly confirmed in their faith; but one wonders, just as the priest who recounted the story perhaps did, just how they might have reacted if they had not got that bear.

Before the missionaries came to Miscou, the "Gougou" terrorized the Aboriginal people there. The Mi'kmaq had told

Champlain that the "Gougou" was an evil monster that dominated the island and the surrounding ocean. It appeared as a woman, they said, a being so huge that the tops of the masts on Champlain's ship would not come up to her waist. Indeed, the whole ship would vanish in her pouch, they said. In this pouch, they recounted, the creature stored the people she caught, and later devoured them at her leisure. Indeed, the "Gougou" ate many people regularly, the Mi'kmaq told Champlain.

This monster was also supposed to make strange, horrible noises near the island. Not only the Aboriginal people, but also Sieur Prévert de Saint-Malo, who had sailed by Miscou with many Mi'kmaq aboard during his search for minerals, told Champlain about passing near the dwelling of this frightful beast, where all those on board his vessel heard the strange, whistling noises that it made. Perhaps they merely heard the bubbling of the unusual freshwater fountain, which Denys discovered welling up in mid-ocean several hundred metres off the island—a fountain which, by the way, no longer exists. [1]

Whatever people saw and heard off Miscou, Champlain seemed unsure. He ended his speculations about the "Gougou" with this reflection: "I conclude that the island is home of some devil who torments them [the Mi'kmaq] in this fashion." What Champlain did not say, and probably did not know, was that Mi'kmaq folklore has always favoured evil giants (*kookwés*), many with cannibalistic propensities. [2] Whether beast or devil, the "Gougou" was not spoken of after the Jesuit occupation of Miscou. There appear to be no accounts of its fearful doings after 1635. Thus, although the mission was not judged a success and was closed in 1662, the missionaries were perhaps more successful than they thought in exorcising the island's demons.

The disappearance of the "Gougou" did not, however, mean the end of strange and unexplained activity in the vicinity of Miscou. The phenomenon of the phantom ship has been closely linked with this island throughout the years. This is understandable since

Miscou is located between two favourite ship haunts, the Bay of Chaleur and the Northumberland Strait. The burning ship, sometimes seen at first as a ball of fire, has been sighted in ancient and modern times, in summer and winter (even on the ice), before, during, and after storms as well as in clear weather, during the day, and at night. Some even say it is an evil omen, "*le feu de mauvais temps*", presaging disaster, as on June 5, 1914, when the ghost ship was seen twice before twenty Miscou fishermen perished at sea during a freak blizzard.

According to one version, Miscou's ghost ship dates from the beginning of the nineteenth century and is associated with the nefarious exploits of a certain Captain Craig who was, we are told, one of the most conniving and pitiless of pirates to pillage coastal settlements around the Bay of Chaleur. Like the earlier piratical tyrant of the inner reaches of the bay, the notorious sixteenth-century Portuguese Captain Corte-Real,[3] Captain Craig's favourite victims were in Mi'kmaq settlements. Both pirates got the Aboriginal people helplessly intoxicated so that they could make off with their most valuable possessions—especially the best animal pelts—and sometimes even the most attractive or prestigious of the Aboriginals.

Captain Craig, like many other captains who entered the Bay of Chaleur, was shrewd enough to realize that in certain parts of the bay he would be safer with a local pilot, familiar with all its idiosyncrasies. Thus, like the merchant captains, he was in the habit of picking up a pilot from Miscou Island before proceeding into the bay. On this particular occasion, the Miscou pilot, following instructions, guided Captain Craig's sailing ship to the Mi'kmaq settlement on the nearby coast, then returned home to wait until he was needed to steer the ship out to sea again.

Several days later, when he had just begun to pilot the boat on its return trip through the most dangerous passage to the open sea, he thought he heard cries and groans coming from the hold. When these noises persisted, he insisted on the sails being lowered so that he could investigate the cause. To his dismay, the pilot found two young Aboriginal girls bound hand and foot. Before agreeing to proceed further, he demanded that they be released. Although the captain and mate were most reluctant to free the girls, they could hardly refuse unless they wished to jeopardize their ship and its precious cargo in such a precarious situation. Thus, the girls were returned to their tribe and the ship began to move seaward once more under the direction of the Miscou pilot. The Aboriginal girls, however, had a premonition that the pirate ship was doomed. Although not displeased that retribution should be meted out to their captors, they suffered agonies at the thought that the man who had saved them would likely perish too.

As the girls had foreseen, the boat sank within minutes of leaving the shore, after striking Miscou Rock. All on board perished except the pilot who, because he was a strong swimmer, reached shore safely. Several months later he married the eldest of the Aboriginal girls, and they are said to have lived a long and happy life together.

From the time of this shipwreck onwards, the pirate ship was often spotted off Miscou. Before its appearance, a powerful red light was seen to detach itself from Miscou Rock and skim over the surface of the bay. Gradually, this ball of fire took on the shape of a boat, which in all its details was said to be clearly recognizable as Captain Craig's boat. It sailed toward shore, and then, within thirty metres of the bank, it disappeared. The comings and goings of the burning ship over the years have been recorded, pondered over, and investigated by local residents and by imported specialists—all to no avail. [4] No one has yet come up with a satisfactory or conclusive explanation.

Settlement on Miscou, as on sister island outposts such as Cape Sable, LaHave, and Scatarie, was attempted at such an early period in this country's history because of the extraordinary promise of its fisheries. For more than four centuries, Miscou islanders have been fishermen first and foremost. Small wonder! In his 1852 *Report on the Sea and River Fisheries of New Brunswick*, Moses Perley lists the varieties of fish found in abundance off Miscou at that time: salmon, halibut, trout, shad, sturgeon, mackerel, cod, ling, flounder, plaice, crab, sculpin, lobster, and shrimp. Schools of some of these fish—such as mackerel—were so prolific that they appeared alongside boats as "almost in a solid mass." [5] Huge numbers of fish were taken by the islanders themselves, by mainlanders, and by visiting American fishermen. Mr. Perley, a conservationist long before such a role was fashionable, was obviously appalled by the widespread waste and mismanagement he observed. Everyone, he indicated, was to blame.

The Jersey houses, which for so many years owned all the fishing stations on Miscou, were troublesome, Mr. Perley stated, because they refused to "furnish salt, even to their best customers to cure herring, mackerel or any pickled fish; and they discouraged the catching and curing of all fish, except such as were dried and fit for the foreign markets." [6] Partly as a result of this practice, most of the fish (particularly herring and mackerel) were used as fertilizer: "It is thought that no more than one tenth of the fish they take are salted and the remaining nine-tenth being put on the fields as manure." [7] Shrimps, brought on shore in large quantities "were turned out again as being of no use." [8] Furthermore, visiting American fishermen were often known to throw out one cargo of fish if a more lucrative one was offered:

> An American vessel fishing off Point Miscou and having on board nearly a full fare of cod, had found the mackerel in such extraordinary abundance, that the crew had thrown overboard one hundred and fifty quintals of green fish, in order to make room for mackerel, with which the vessel had been quickly filled up. [9]

Yet despite such waste, the Miscou fisheries were not noticeably depleted by the next century; 1916 was a record year. That was the year when a good fisherman could take in twenty-two hundred pounds of lobster in a single trip! (His pay for such a catch was a cent and one half a pound.) Miscou's population has always depended on the ups and downs in the fisheries. When the fisheries thrived and the markets were strong, the population grew; when the catches diminished or the markets decreased, an exodus of island dwellers was sure to result. Because settlements on this island were started so often over the years and then abandoned, paradoxically, the permanent occupation (begun about 1815) was much later than on such Maritime islands as Brier, Grand Manan, Tancook, and others that were settled for the first time in the late-eighteenth century but were then tenaciously held onto by the descendants of these early immigrants.

Another paradoxical fact about Miscou's final settlement is that it was begun by anglophones. Given its location and early history, one is at first greatly surprised to learn from so reliable a source as Ganong's *History of Miscou* that "of all the settlements of the North Shore of New Brunswick, this one alone owes the inauguration of its permanent settlement to the English rather than to the French." The place-names—"Wilson's Point," "Miscou Harbour," and "Miscou Point"—attest to the continuing influence of the first anglophone residents, as do the small Protestant churches in both Miscou Harbour and Miscou Point.

However, in the last bid for settlement on Miscou, francophone settlers soon arrived from many parts of "Acadie," from France, from Quebec, and from the Island of Jersey. They put down roots in Grande Plaine, a village that had once before (from about 1719 to 1730) gained prominence in this island's history because of the amazing numbers of sea cows or walruses slaughtered there. The rapid extermination of these gentle creatures coincided with the demise of the settlement.[10]

The new francophone settlers at Grande Plaine were fishermen like most of the earliest island residents and like their anglophone neighbours. This final settlement at Grande Plaine flourished and grew so that soon several families moved inland to establish the community, which is now known as Miscou Centre. Over the years, the francophone population has grown more quickly than the anglophone, and now most of the island's population is French and is concentrated in Miscou Centre where there is a large Catholic church, a school, and several stores. The number of people on the island has fluctuated during this century as much as in the three preceding centuries. In 1906 W. F. Ganong recorded the island's population as between four and five hundred. This figure, he predicted (probably correctly) was the limit the land and surrounding sea could comfortably maintain. Yet, because of a thriving fishery, the population grew until, by 1947, it was three times Ganong's ideal number—about 1,500. After this, with the closure of many fish plants, the numbers dwindled.

Like so many other offshore islands of the Maritime provinces, Miscou was to some extent affected by the general twentieth-century trend away from island living—a trend that peaked in the early fifties. Whereas in this rapid exodus, such long-established island communities as those on Scatarie, Cape Negro, Cocagne, Heron, Shediac, Liscomb, and McNutt's were deserted almost overnight, Miscou's colony shrank only gradually, and only toward Ganong's ideal population figure. By 1973 there were half as many residents as there had been in 1947: fifty-two families (745 people).

According to existing accounts, most of those who now remained on the island were unlikely to be dislodged.[11] They were tenacious in clinging to their island roots. One old man is supposed to have announced, "They won't get me out of here alive and make me go and live somewhere else in a shack. Deprived of the Gulf air, it would be better to die!" And down the road in Miscou Centre a ninety-year-old woman (in the spring of 1973) was out digging her garden, her left arm in a sling because she had broken her wrist toward the end of the winter. This woman, Annie Ward, was obviously cut from the same strong fabric as Christiania Davis, the Brier Island woman who had walked to Halifax and back to settle a land dispute.

Miscou residents somehow managed to live with the main problem faced by all Northumberland Strait or Bay of Chaleur island residents—namely, unsafe ice (particularly during freeze-up and thaw)—which can make crossing to the mainland a hazardous undertaking. They obviously felt that their lifestyle was worth the risk. "A causeway from Miscou to Lameque would change the pace of life here, in the winter especially," one man remarked. "It wouldn't be so tranquil here if this wasn't a real island. As things are now, you have to think ahead and lay in supplies for winter. Planning that far ahead makes for a different mentality than most people on the mainland have now."

But whether French or English, Catholic or Protestant, the island residents have seemed happy with their lot and, on the whole, pleased with one another. As an eighth-generation descendant of a Jersey fisherman remarked: "We are one big, happy family here. We all love the island and the sort of life it gives us. You must remember that living on an island means that everyone *has* to look out for his neighbour. That improves the quality of life!"

Notes on Miscou Island

1. In 1906 W. F. Ganong went to the island and searched for this fountain, but was unable to find it.
2. See Rev. Silas T. Rand, *Legends of the Micmacs,* One tale of a cannibalistic giant included in this book is "The History of Kitpooseaguno—A Tale of Ancient Times" (62–80). Rand carefully transcribed such tales as this from the original oral tales. Rand was, by the way, a painstaking scholar and an extraordinary linguist who spent forty years working among the Mi'kmaq studying their language and culture—in a comparative context, it might be added. In the course of his studies, Rand mastered twelve languages (English, Latin, Greek, Hebrew, French, Italian, German, Spanish, Modern Greek, Mi'kmaq, Maliseet, and Mohawk) and acquired great insight into their respective cultures.

3. For an account of Corte-Real's doings see chapter on Heron Island.
4. The most extensive listing of material relating to the sightings and explanations (both scientific and folkloric) is in *Le Vaisseau Fantome* (*Légende étiologique*) by Soeur Catherine Jolicoeur.
5. Moses Perley, *Report on the Sea and River Fisheries of New Brunswick* (36). Even quite recently, Miscou had an unusual reputation connected with its abundant seafood. A brief commentary by Michael Camp in the *Telegraph-Journal* (January 31, 1984, p. 5) begins as follows: "If a strong northeast wind suddenly shifts to the northwest, chances are the current will wash hundreds of lobsters onto the beaches of this small island. If you're quick enough you can collect a pail-full or two for dinner."
6. *Ibid.*, p. 34. Perley also quotes a Mr. Wilson of the island, who states: "...the Jersey houses exact too large profits, and keep the fishers in poverty; they look only to one branch of fishing and discourage all others" (36). Elsewhere in his report, Perley refers to the "absolute state of serfdom" of the Point Miscou fishermen because of the practices of the Jersey houses (p. 39).
7. *Ibid.*, (34).
8. *Ibid.*, (37).
9. *Ibid.*, (36).
10. For a touching account of this period in Miscou's history see Louis Haché's novel, *Adieu P'tit Chipagan*.
11. See Louis Haché, *Charmante Miscou* (27; 29). "On ne me sortira pas d'icitte vivant pour aller vivre ailleurs dans un shack, proteste un vieux Miscou. Privé de l'air du golfe, mieux vaut mourir!" (27) Michael Camp in the *Telegraph-Journal* (January 31, 1984, p. 5) lists the population of Miscou Island as 1,100.

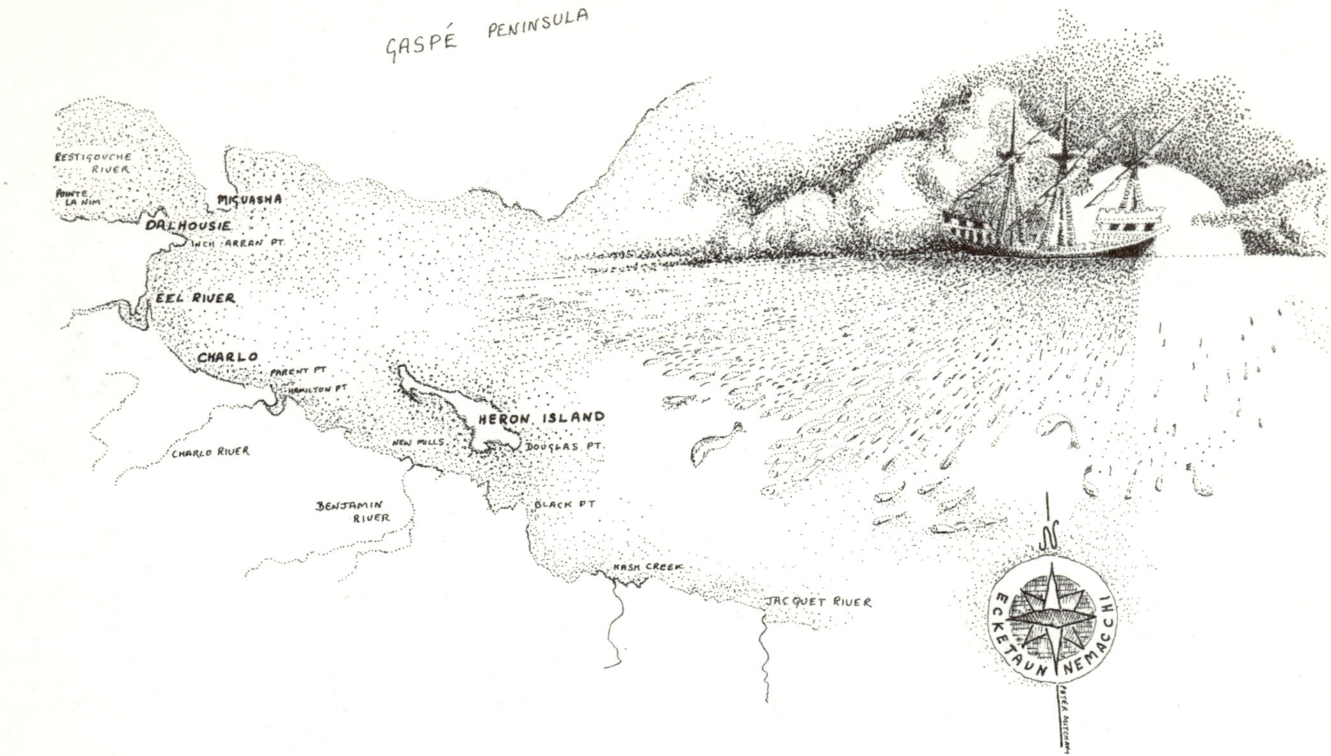

GASPÉ PENINSULA
RESTIGOUCHE RIVER
POINTE LA NIM
MIGUASHA
DALHOUSIE
INCH ARRAN PT.
EEL RIVER
CHARLO
PARENT PT
HAMILTON PT
CHARLO RIVER
HERON ISLAND
NEW MILLS
DOUGLAS PT.
BENJAMIN RIVER
BLACK PT
NASH CREEK
JACQUET RIVER
N
ECKETAUN NEMACCHI
PETER MITCHAM

Heron Island

The Bay of Chaleur possesses many advantages for the prosecution of the fisheries. The whole Bay may be considered one great harbour, as throughout its entire breadth and extent there is not a single rock, reef or shoal. During the summer it literally swarms with fish of every description known on the shores of British North America: and its ancient Indian name of Ecketaun Nemacchi—The Sea of Fish—well denotes its character.

—Moses Perley (1852) *Report on the Sea and River Fisheries of New Brunswick*

The most significant island in this "Sea of Fish" is Heron Island, which lies between New Brunswick and the Gaspé coast where the Bay of Chaleur begins to narrow to meet the Restigouche River. This long and narrow island (about eight kilometres by one and a half kilometres, or less) is roughly three kilometres off the New Brunswick coast and more than triple that distance from Quebec. Because of its protected location and excellent fishing—comparable for these reasons to the Northumberland Strait islands of Cocagne and Shediac—it seems an ideal habitat for herons, who appear from earliest recorded times to have been associated with this island. The Mi'kmaq, too, found this place greatly to their liking, as did white immigrants who eventually, like the herons and the Aboriginal people, settled here because of the excellent fishing and the shelter it afforded.

Settlers or visitors, whether French or English, have consistently kept the same name for the island. In the earliest Louisbourg records, frequent brief references are made to voyages undertaken to "La Baie des Chaleurs," where "l'île-aux Herons" was sighted; and late eighteenth-century accounts use the same spelling as the Louisbourg records when referring to the granting of this island, together with some of the nearby mainland—including the land on which the town of Charlo now stands—to l'Abbé Bourg, the first Acadian priest serving as missionary in the Bay of Chaleur region.

L'Abbé Bourg had this land pressed upon him in the 1780s as a reward for quelling an Aboriginal uprising. (Apparently only his timely intervention saved bloodshed.) Several sources indicate that no other white man of this time possessed so extensive a knowledge of Aboriginal languages and customs or was able to establish so strong a rapport with the native people as l'Abbé Bourg. Ironically, however, although Sir Richard Hughes made a great fanfare—which resounded from Hughes' own residence in Halifax to the good Abbe's residence at Carleton (on the Gaspé across from Heron Island)—trumpeting Bourg's accomplishment and his own delight in rewarding him, he never actually gave Bourg a deed to the land that he seemed so eager to confer on him. Bourg, moreover, who was by all accounts an unworldly man, appears to have been so concerned with attending to the spiritual welfare of his own people and the Mi'kmaq under his jurisdiction that he had no time or energy to attend to lands on Heron Island.

It was not, therefore, until almost mid-nineteenth century that Heron Island was divided into twelve farms of unequal acreages, stretching from shore to shore (north to south). These grants were taken up by anglophone families whose chief occupation was fishing. As Moses Perley pointed out in his 1852 report: "There is excellent herring fishing around this island...lobster and sea trout are abundant." Nevertheless, Mr. Perley was upset by some of the inequalities governing the fishing here even at such an early date. The best fishing, he noted, was between the island and the New Brunswick mainland, and the lots already granted covered all the beaches, leaving no fishing stations available for newcomers.

Mr. Perley also quoted remarks made by a nearby mainland resident—a Mr. Harvey of Nash's Creek—that Americans should be allowed to "prosecute the fisheries [here], as they would teach the young men the latest and most approved modes of fishing, from ignorance of which they could not at present follow fishing profitably." That Mr. Perley did not protest against this viewpoint is surprising, given his knowledge of the difficulties resident fishermen of both Grand Manan and Miscou were having at this time with American interlopers and his perspicacity in spotting and calling attention to troublesome situations in advance.[1]

The American boats that fished so close to Miscou at this time, and periodically entered the Bay of Chaleur, were chiefly after cod, since the cod in the bay was reputed to be of superior quality for shipping to the Mediterranean markets, where it was much in demand. The Bay of Chaleur codfish were smaller than those caught elsewhere and were said to remain white and delicate even after being salted and dried.

Apart from its proximity to fine fishing, Heron Island has never had much in common with the larger neighbouring island at the mouth of the bay. On windswept Miscou, sparse evergreens struggle valiantly to hold their ground; on sheltered Heron, lush stands of cedars, poplars, birches, maples, as well as spruce and fir, reach upward and outward as they vie with one another for a place in the sun. Miscou's many bogs produce ground-hugging vegetation—pitcher plants, cranberries, and bakeapples—Heron's well-drained upland meadows generate shoulder-high grasses, multitudes of wildflowers, and raspberry bushes.

Heron Island, like Merigomish Island in Pictou County or Indian Island near Richibucto, was, understandably, a favourite haunt of the Mi'kmaq. Here they dwelt under seemingly ideal conditions until the coming of the white man caused them to abandon their beautiful island forever. The repercussions from this first clash between Aboriginal people and European mariners proved disastrous for all concerned. According to the story, the tragedy began in 1500 when an unscrupulous Portuguese adventurer, Gaspar Corte-Real, arrived in the Bay of Chaleur. Going ashore on the Gaspé coast opposite Heron Island, he and his crew plied a number of the chiefs with alcohol until they were hopelessly drunk, then took them prisoner and transported them to Europe to sell as slaves—apparently not at all concerned that this treachery might boomerang.

Corte-Real returned to the Bay of Chaleur the following summer, this time to Heron Island. Since the Mi'kmaq there seemed friendly, he assumed that either they knew nothing of the kidnapping that had taken place on the nearby Gaspé coast the previous summer, or that if they knew, they did not care. Thus Corte-Real and his men, relaxing all vigilance, established themselves on the island and focused their attentions on trading with the Mi'kmaq. The Mi'kmaq, however, were both cannier and more unforgiving than they seemed; they were merely biding their time for vengeance.

One dark night, when the Portuguese were asleep, the Mi'kmaq stole into the Europeans' dwelling, murdering everyone except Corte-Real. For him, they had devised a punishment in keeping with the crime he had committed against their people the preceding summer. Taking Corte-Real to a rock, which was not submerged at low tide, they secured him to it and then abandoned him so that he would have a number of hours to reflect on his wicked life and approaching death before the tide rose over him.

Gaspar Corte-Real's death did not, however, put an end to unpleasant foreign intrusions (as no doubt the Mi'kmaq had hoped), for the following summer (1502) Gaspar's brother, Miguel Corte-Real, came to find out what had happened to the captain and crew of the first trading vessel. Almost at once he spotted the *Caravel*

The death of Gaspar Corte-Real

moored, still intact, off the middle part of the island. As he saw no sign of life aboard, he sailed his own ship closer. Suddenly a large number of Aboriginal warriors shot out from shore in their canoes, clambered quickly aboard the vessel, and at once massacred everyone in sight.

Seeing the fate of their fellows, the captain and surviving crew barricaded themselves in the cabin and below decks with all the weapons they could lay hands on. However, the *Caravel*, left untended, began to drift out to sea, and the sailors surmised that they were doomed. Before making a final rush on the Mi'kmaq on deck, the Portuguese are supposed to have said a prayer which included the vengeful threat that if they died, they would return to haunt the Bay of Chaleur for one thousand years.

During the final confrontation between the Portuguese and the Mi'kmaq, the boat caught fire. Soon the sails and masts were ablaze and the gutted boat disappeared quickly beneath the waters of the bay. Only one Aboriginal warrior survived to tell the story.

We were told that even after all this the Mi'kmaq were not left in peace, because from this time on, they kept seeing the burning ship bearing down upon them—particularly on the eve of a storm. Increasingly terrified by these strange and frequent reappearances of the lost ship, they abandoned Heron Island and went to live on the banks of the Restigouche River.

Ever since this time the burning ship has been spotted sailing up and down the bay in all seasons of the year, though it is most often visible, some say, just before a storm. Hundreds of men, women, and children have reported watching the vessel burn and, at length, sink into the sea. [2] Many contemporary viewers seem as puzzled and upset by the phenomenon as the Mi'kmaq are supposed to have been nearly five centuries ago. The settlers who finally became established on Heron Island, however, seem to have been little troubled by ghosts of the past. Neither the restless spirits of the displaced Mi'kmaq nor the vengeful specters of the murdered Portuguese returned to mar the tranquility of their lives or the abundance of the harvests from land and sea. As former residents have been quick to emphasize: "Life on Heron Island was idyllic."

One man, Stanley LaPointe, who took his family back every summer to camp on the beach below his family's former farmhouse, remarked:

> Nowhere else has the same feeling of peace and plenty for me.... Even in the Depression with eleven kids in our family we always had enough. Like everyone else on the island, we had a good house, some animals, a great garden, berries galore, and lots of

> fish and firewood. Of course my father seemed almost rich at the time because, in addition to everything else that I have mentioned, he had two concurrent government jobs—lighthouse keeper and postmaster. Yes, we were lucky. But now it's best that no one lives here. I'm glad the island is deserted. This way there's no way of it getting spoiled.

Mr. LaPointe did not elaborate on this last remark, but one only needs to see what has happened to nearby parts of the mainland—at Belledune, for example—to know what he meant. Not so long ago that was once a lovely spot too. Because the whole bay is, as Mr. Perley noted, such a fine and safe natural harbour of immense proportions, industries have seen the immediate advantages of locating here. By putting up ugly, sprawling, concrete buildings, crowned with immense smokestacks and flanked by black-topped and machine-littered acres, they have already marred the scenery of one of the world's great natural beauty spots. Heron Island has fortunately been spared this sort of progress.

However, like the rest of the bay, the waters around Heron Island are slowly but surely being polluted by the many chemicals that industries all along the bay spew daily into the once-pure waters. Not only are the fish less numerous than they once were, but questions have been raised about the toxic levels of contaminants in the survivors. [3] Unless we can somehow manage to alter aspects of our development, "Echetaun Nemaachi," or "The Sea of Fish," will be nothing but a vague and haunting memory.

Notes on Heron Island

1. For a detailed account of the problems that the Miscou fishermen faced, see Perley's 1852 *Report on the Sea and River Fisheries of New Brunswick,* and, for a record of the difficulties in the Grand Manan

fisheries, see Perley's 1850 *Report on the Fisheries of Grand Manan.* Actually, American boats had been coming to the Bay of Chaleur for the best part of the hundred years before Perley made his observations and wrote his report in 1852. This "Sea of Fish" was for years the primary destination of New England fishing boats and Jersey merchants alike. One of these Jersey merchants, Charles Robin, in his *Journal* (much of which is a day to day account of his business trips from Arichat, Isle Madame, Cape Breton to the Bay of Chaleur during the years 1767–1769) tells of the many schooners and whalers he observed each spring anchored between Canso and Isle Madame waiting for the drift ice to shift so that they could pass through into the Northumberland Strait. Once into the strait, they steered for Pictou Island, much as earlier fishermen along the Cape Breton coast had been guided by Scatarie Island as a landmark. However, unlike their European predecessors on the Cape Breton coast, there are no records of this new wave of transient fishermen either stopping on the island or coming to grief on its coast. Intent on reaching the "Sea of Fish," the only impediments once the ice had gone from the usually tranquil Northumberland Strait were sandbars off Cape Tormentine. Thus, reaching their favourite fishing grounds was usually neither a difficult nor dangerous undertaking.

2. The greatest number of accounts of these sightings are brought together in Soeur Catherine Jolicoeur's book, *Le Vaisseau fantôme, légend étiologique.*
3. See Warner Troyer's *No Safe Place.*

Source Material

Below are listed some of the works consulted while researching this book. This list, however, represents only a small portion of the source material. For instance, particularly useful were manuscripts, letters, census reports, and maps available in various archives and libraries such as: Les Archives acadiennes (Université de Moncton), the Provincial Archives (Halifax and Fredericton), the Public Archives (Ottawa), the Louisbourg Archives, the Mount Allison Library, and the Legislative Library (Fredericton). Much helpful information has also been provided by the Grand Manan Historical Society. As important as any of the above is all the information collected by word of mouth from scores of former island residents; from lighthouse keepers and their families, and from fishermen—particularly those who provided transportation to the various islands which are the subject of this book.

Selected Bibliography

Barto, Martha Ford. *Genealogies of West Isles Families*. Saint John: Lingley Printing Co., 1975.

Boyd, James. *Diary of W. H. Kilby 1763*. Eastport and Passamaquoddy: n.p., 1888.

Cameron, Dr. Ian. "Cholera Threatened City." *Nova Scotian*. August 24, 1985 (3).

Chabert, M. de. *Voyage fait par ordre du roi en 1750 et 1751 dans l'Amérique Septentrionale*. New York: Johnson Reprint Corporation, Mouton and Co., 1966.

Champlain, Samuel de. *The Works of Samuel de Champlain*. 6 vols. Toronto: The Champlain Society, 1922.

Clos, M. "Memorial on behalf of the Sieur de Boishébert, captain, chevalier de Saint-Louis, former commandant in Acadia." Trans: Louise Manny. Ed: Dr. J. C. Webster. *Historic Studies*, no. 4. Saint John: New Brunswick Museum, 1942.

Cooney, Robert. *A Compendious History of the Northern Part of the Province of New Brunswick*. Halifax: Joseph Howe, 1832.

Cushing, J. Elizabeth; Theresa Casey; Monica Robertson. *A Chronicle of Irish Emigration to Saint John, New Brunswick (1847)*. Saint John: New Brunswick Museum, 1979.

Daigle, L. Cyriaque. *History de Saint-Louis de Kent*. Moncton: l'Imprimerie Acadienne Ltd., 1948.

Denys, Nicholas. *Sa vie et son oeuvre*. Yarmouth: L'Imprimerie Lescarbot Ltée, 1982.

---. *The Description and Natural History of the Coasts of North America (Acadia)*. Toronto: The Champlain Society, 1908.

Faucher de Saint Maurice, Henri-Edouard. *Promenades dans le Golfe Saint-Laurent*. Montreal: Cadieux & Derome, 1881.

Ganong, W. F. "A monograph of the place-nomenclature of the province of New Brunswick." Transactions of the Royal Society of Canada (3, 2, 2), 1986.

---."A Monograph on the Origins and Settlements in the Province of New Brunswick." *Contributions to the History of New Brunswick*, no. 6, 1904.

---."Champlain's Island; Ste. Croix (Dochet) Island." *Museum Monographic Series*, no. 3. Saint John: New Brunswick, 1945.

---.*Historic Sites in the Province of New Brunswick*. St. Stephen: Print 'n Press Ltd., 1983.

---."The History of Miscou." *Acadiensis*. Fredericton: University of New Brunswick, 1906.

---."The Physical Geography of the North Shore Sand Islands." *Bulletin of the Natural History Society of New Brunswick*, vol. 6, no. 26, n.d.

Godfrey, William. *History of Chatham*. Chatham: Miramichi Historical Society, 1962.

Grand Manan Historian, vols. I–XXI. Grand Manan: Grand Manan Historical Society.

Haché, Louis. *Charmante Miscou*. Moncton: Editions d'Acadie Ltée., 1974.

---.*Adieu, P'tit Chipagan*. Moncton: Editions d'Acadie Ltée., 1979.

Haliburton, Thomas Chandler. *An Historical and Statistical Account of Nova Scotia*, vols. I–II. Halifax: J. Howe, 1829.

Hamilton, W. O. and W. A. Spray. *Source Materials Relating to the New Brunswick Indians*. Fredericton: Hamray Books, 1977.

Jolicoeur, Sr Catherine. *Le Vaisseau fantôme, légende étiologique*. Quebec: Les Presses de l'Université Laval, 1970.

Le Centre d'études acadiennes. *Inventaire General des Sources Documentaire sur les Acadiens*. Moncton: Editions d'Acadie, 1975.

Lorimer, John G. *History of the Islands and Islets in the Bay of Fundy*. St. Stephen: Saint Croix Courier, 1876.

Losier, M. J. and C. Pinet. *Children of Lazarus*. Fredericton: Fiddlehead, 1984.

MacNutt, W. S. *New Brunswick: A History 1784–1867*. Toronto: MacMillan, 1984.

Martin, Lois. *Historical Sketches of the Miramichi*. Chatham: Bicentennial Committees of Chatham and Newcastle, 1985.

Melançon, Abbé Arthur. *Vie de l'Abbé Bourg, premier prêtre acadien*. Rimouski: Le "Chez Nous," 1921.

Mersereau, Rev. Charles. "Beaubears Island." n.p., n.d.

Milner, W. C. and H. Robinson. *History of Albert County*, 2 vols. Saint John: New Brunswick Museum, 1934.

Mitcham, Allison. *Abraham Gesner: Prophet of the Wilderness*. Hantsport: Lancelot Press, 1995.

———. *Three Remarkable Maritimers*. Hantsport: Lancelot Press, 1985.

———, ed. *The Best of Abraham Gesner*. Hantsport: Lancelot Press, 1995.

Mitchell, John. *Field Book of Surveys in Passamaquoddy*. n.p., n.d.

Murdoch, Beamish. *A History of Nova Scotia or Acadie*. Halifax: James Barnes. Vol. I (1865), Vol. II (1966).

Perley, Moses. *Report on the Sea and River Fisheries of New Brunswick*. Fredericton: Queen's Printer, 1852.

———. *A Handbook of Information for Emigrants to New Brunswick*. Saint John: H. Chubb and Co., 1854.

Rand, Silas T. *Legends of the Micmacs*. New York: Longmans, Green & Co., 1894.

Rayburn, Alan. *Geographical Place Names of New Brunswick*. Ottawa: Department of Energy, Mines, and Resources, 1975.

Raymond, W. O. *Winslow Papers, A. D. 1776–1826*. Saint John: New Brunswick Historical Society, 1901.

Relations des Jésuits. (Tome 4 [1647–1655)]). Montreal: Editions du Jour, 1972.

Robin, Charles. *Journal of Charles Robin (1767-1787)*. Ottawa: Public Archives, n.d.

Savoie, Francis. *L'Ile de Lamèque*. Moncton: Editions d'Acadie Ltée, 1981.

Smethurst, Gamaliel. "A Narrative of an Extraordinary Escape out of the Hands of the Indians in the Gulf of St. Lawrence." London: n.p., 1774.

Stanley, Laurie C. C. *Unclean! Unclean! Leprosy in New Brunswick 1844–1880*. Moncton: Editions de l'Acadie, 1982.

Still the Gentle Breezes Blow: Essays on the History of Harvey Parish, Albert County, New Brunswick. Albert: Harvey Parish Bicentennial committee, 1984.

Surrette, Paul. *Le Grand Petcoudiac*. La ville de Dieppe, n.p., 1985.

Troyer, Warner. *No Safe Place*. Toronto: Clarke, Irwin, 1977.

Wilson, Beckles. *The Life and Letters of James Wolfe*. London: Heinemann, 1909.

Wright, Esther Clark. *The Miramichi*. Sackville: The Tribune Press, 1944.

———. *The Peticodiac*. Sackville: The Tribune Press, 1945.

Wright, Harold, ed. *The Diary of Nellie McGowan, Partridge Island Quarantine Station, 1902*. St. John: Patridge Island Research Project, 1984.